安徽省高水平高职教材
高职高专化工类系列教材

化学化工专业英语教程

ENGLISH FOR CHEMISTRY AND CHEMICAL ENGINEERING

主　　编　吕丹亚　樊陈莉

副主编　潘　莉

编写人员（以姓氏笔画为序）

韦琛鸿　刘义章　吕丹亚
吴　康　周韦明　洪　娟
董　泓　潘　莉　樊陈莉

中国科学技术大学出版社

内 容 简 介

本书为安徽省高水平高职教材，从化学化工专业英语的基本特点出发，打破传统的“课文+练习”的编写方式，采用专题模式，介绍了化学化工英语的词汇、翻译与阅读、专业写作，具有专业性强、实用性强、针对性强、难易程度适中的特点，可作为高职高专院校化学化工类专业学生的专业英语教材，也可供化学化工专业的从业人员参考。

图书在版编目(CIP)数据

化学化工专业英语教程/吕丹亚，樊陈莉主编.—合肥:中国科学技术大学出版社，2021.8
ISBN 978-7-312-05205-7

Ⅰ.化… Ⅱ.① 吕… ② 樊… Ⅲ.① 化学—英语—高等学校—教材 ② 化学工业—英语—高等学校—教材 Ⅳ.① O6 ② TQ

中国版本图书馆CIP数据核字(2021)第067160号

化学化工专业英语教程
HUAXUE HUAGONG ZHUANYE YINGYU JIAOCHENG

出版 中国科学技术大学出版社
安徽省合肥市金寨路96号，230026
http://press.ustc.edu.cn
https://zgkxjsdxcbs.tmall.com
印刷 安徽省瑞隆印务有限公司
发行 中国科学技术大学出版社
经销 全国新华书店
开本 787 mm×1092 mm 1/16
印张 14.5
字数 371千
版次 2021年8月第1版
印次 2021年8月第1次印刷
定价 45.00元

前　　言

本书从化学化工专业英语的基本特点出发，内容涵盖化学化工英语词汇、翻译与阅读、专业写作等，共11章。本书中英文结合，讲解部分以中文为主，练习部分以英文为主，具有专业性强、实用性强、针对性强、难易程度适中的特点。

绪论部分以化学化工英语为主，介绍了科技英语的特点，让学生对科技英语的语法、词汇和写作有一个初步的了解。如：在语法方面，科技英语文章多用被动句、名词化结构和复杂的长句，语言精练、结构严谨。其词汇构成的主要方式为合成词、派生词和缩略词等。知识点辅以具体的例子进行说明，便于学生掌握。

在化工专业英语词汇部分，本书对化学化工英语的词汇特点和规律作了总结，便于学生理解记忆；在无机化学部分，介绍了无机化合物的命名方法，包括阳离子、阴离子、酸、碱、盐、分子化合物和水合物的命名规律，并辅以范例说明；在有机化学部分，介绍了包括烷烃、烯烃、炔烃、环烃、芳烃、醇、醚、酚、醛、酮、羧酸及其衍生物等有机化合物的构成和命名特点。

翻译与阅读部分针对化学化工英语的特点、翻译标准和翻译方法，重点介绍了直译和意译、词性转换译法和增词、省词译法，选编了名词性从句、定语从句、状语从句和被动语态等译法，并提供丰富的例句供学生学习。阅读部分的材料注重从化工的角度选择专业阅读材料，所选文章涉及的工艺为化学化工领域最具代表性的工艺，涉及无机化学、有机化学、分析化学、物理化学、生物化学、化学工程等，便于学生扩大知识面和掌握更多的专业词汇。本书通过对文章中的长句进行语法分析，以理清句子成分，教会学生翻译技巧，使其能用准确、简洁和规范的汉语表述原文。阅读与翻译部分覆盖面广、内容丰富，便于不同地区的院校根据化学化工专业开设的学时数和研究方向选学不同章节的内容。

专业英语写作部分介绍了化学化工英语论文的结构、摘要的撰写、文献检索、文献综述的写作，旨在帮助学生能更快掌握行业资讯，了解前沿科技产品及工艺；还介绍了化学化工英语论文写作的类型及文体特点、语句表达的规范、摘要的写作及如何发表论文等知识，帮助学生更好地完成毕业设计和论文。写作部分介绍了论文撰写的格式，即题目、作者、附属机构（即作者所在单位）、摘要、关键词、目录、前言、正文、方法、结果、讨论、结论、致谢、参考文献和附录等，重点介绍了论文摘要的撰写、综述的写作要求以及常用的专业写作句型。

本书可作为高校化学化工类专业学生的专业英语教材，也可供化学化工专业的从业人员参考。

本书在“校企合作、双元开发”的基础上由芜湖职业技术学院吕丹亚、樊陈莉担任主编；芜湖职业技术学院潘莉担任副主编；安徽神剑新材料股份有限公司周韦明，安徽国星生物化学有限公司韦琛鸿，滁州职业技术学院刘义章，安庆职业技术学院董泓，芜湖职业技术学院洪娟、吴康参与编写。全书由吕丹亚、洪娟统稿。本书在编写过程中得到了芜湖职业技术学院有关领导的帮助和指导，在此表示感谢。此外，本书也得到了芜湖职业技术学院学生孙见宾、谢子龙、周成威、单厚林等的支持。

由于编者水平有限，加之时间仓促，存在疏漏和不妥之处在所难免，恳请读者提出宝贵意见，以便完善。

编　者

2021年5月

目　　录

第2篇 翻译与阅读

第3篇 专业英语写作

绪　论

科技英语(English for science and technology,简称EST)一般指在自然科学和工程技术方面的科学著作、论文、教科书、科技报告和学术演讲中所使用的英语。科技英语诞生于20世纪50年代,在第二次世界大战后得到迅速发展并受到学术界广泛的关注和研究。

化学化工专业英语属于科技英语,是经过长期发展形成的化工行业英语,是旨在用英语阐述化工行业中的理论、技术、实验和现象等的英语语言体系,在词汇、文体和语法诸多方面都有自己的特点。

0.1 词汇特点

1. 词义专一

化学化工专业英语词汇的词义相对专一,不像文学作品中的英语词汇常出现一词多义或一义多词的现象。在表达同一个科学概念或含义时,一般采用单一词汇,基本上不会出现歧义。例如:

methane　甲烷　　　　rectification　精馏

polypropylene　聚丙烯　　　　salicylic acid　水杨酸

2. 广泛使用缩略词

专业英语词汇广泛采用缩略词,便于记忆和书写,因而使用频率高。缩略词可以是每个单词的第一个字母,如高效液相色谱(HPLC, high efficiency liquid chromatography)、聚氯乙烯(PVC, polyvinyl chloride);也可以是单词的部分字母,如氯氟烃(CFCs, chlorofluorocarbon);还可以是一个单词的首字母加上另一个单词,如氢弹(H-bomb, hydrogen bomb)。

有的单词无论是拼写还是发音都像是一个完整单词,但其实也是缩略词,如激光(laser, light amplification by stimulated emission of radiation)、雷达(radar, radio detection and ranging)等。

3. 派生词出现的频率很高

英语的构词法主要有3种:合成法、转化法和派生法。其中,很多科技英语词汇都是由派生法形成的,即通过添加前缀或后缀,构成新的单词,如:“micro-”表示“微小的”,由前缀

“micro-”构成的化工英语词汇有：

microorganism 微生物　　micromole 微摩尔

microwave 微波

再比如：“anti-”这个前缀可以表示“抗”“防”和“反对”的意思，通过添加“anti-”为前缀构成的化工英语词汇有：

antibacterial 抗菌的　　antibody 抗体

antifreeze 防冻剂　　antibiotic 抗生素

同样，借助后缀也可以构成化工英语词汇，如后缀“-ane”表示“烷烃”，结合科技英语中表示“一”到“十”的前缀，可以构成甲烷等烷烃的词汇，如：

甲烷 methane　　己烷 hexane

乙烷 ethane　　庚烷 heptane

丙烷 propane　　辛烷 octane

丁烷 butane　　壬烷 nonane

戊烷 pentane　　癸烷 decane

此外，在科技英语中，还大量使用逻辑语法词，分类如表0.1所示。

Table 0.1 Logical gramnar vocabulary

表0.1 逻辑语法词

功能	逻辑语法词
表示原因	because, as, for, because of, caused by, due to, owing to, as a result of
表示转折	but, yet, however, otherwise, nevertheless
表示逻辑顺序	so, thus, therefore, furthermore, moreover, in addition to
表示限制	only, if only, except, besides, unless
表示假设	suppose, supposing, assuming, provided, providing

0.2 文体特点

化学化工专业英语文体质朴、语言精练、结构严谨，避免使用晦涩难懂的表达方式，极少运用修辞手法，以体现内容的准确性、连贯性和清晰性。描述科学的语言注重事实和逻辑，因此往往以图表、公式、数字来表达科学概念，普遍使用逻辑语法词，如下段介绍吸收实验流程的英文：

Figure 0.1 presents a diagram of the gas absorption unit to be comployed in this experiment. This unit consists of 1 air compressor, 2 Raw material storage tank, 3 rotameters, 1 heater, 3 thermometers and an absorption tower. An air-acctone mixture is introduced to the bottom of the column and pure water is sprayed from the top of the column. The air flowing upward contacts with water stream flowing downwards, which absorbs acetone.

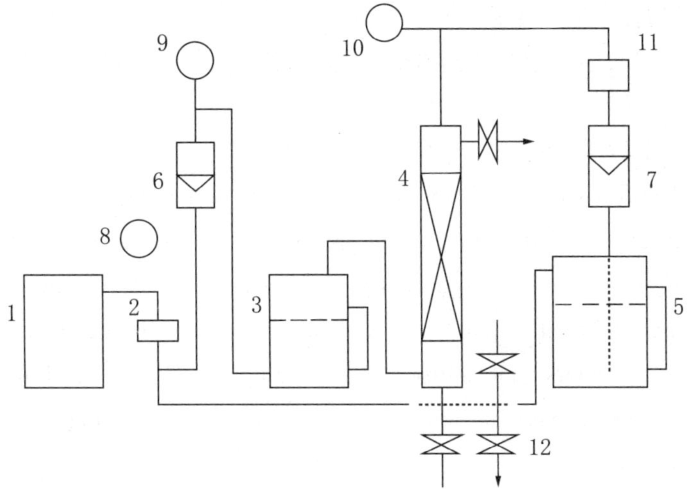

Figure 0.1　The absorption experiment flow chart

图0.1　吸收实验流程图

1. Air compersso r; 2. Pressure; 3. Acetone bubbler; 4. Packing tower; 5. Water tank; 6. Gas rotameter; 7. Liquid rotameter; 8. Pressure gage; 9、10. Thermometer; 11. Electric heater; 12. Liquid sealing device.

1. 空气压缩机; 2.压力调节器; 3.丙酮鼓泡器; 4.填料塔; 5.水箱; 6.气体转子流量计; 7.液体转子流量计; 8.压力表; 9、10.温度计; 11.电加热器; 12.液封装置。

0.3　语法特点

化学化工专业英语具有鲜明的语法特征,主要表现为大量使用长句、被动句、名词化结构、后置定语、非谓语动词等。

1. 大量使用长句

化工英语文章要求叙述准确,逻辑严谨。因此,为了清楚地表达一个复杂的概念、事实或者实验过程,往往使用长句。长句结构严密,句子成分相对复杂。

例如:

Drexler imagines that each of these assemblers is a micro-scale robot with nanoscopic manipulator arms that are capable of picking up individual atoms and sticking them together in any fashion programmed in its internal memory.

在这个长句中,句子主干是"Drexler imagines that..."(德雷克斯勒设想……),that 引导的宾语从句是设想的内容。这个宾语从句是主语+系动词+表语结构,即"Each of these assemblers is a micro-scale robot."(每一个万能装配手都是一个微米大小的机器人)。"with nanoscopic manipulator arms"(带有纳米尺寸的操作臂)是介词短语做后置定语,其中,"arms"后接定语从句,描述其作用是"that are capable of picking up individual atoms and sticking them together in any fashion programmed in its internal memory."(能够抓起单个的原子并按照

内部存储器里设定的程序方式将它们粘在一起)。

虽然这个句子很长,但由于其结构清晰、逻辑性强,分析句子各成分之间的关系后,便很容易理解句意。

再如,下面这一个句子中就有76个单词:

They may be found working in industrial laboratories and production departments, private and government institutes, and the laboratories of colleges and universities, chemists and other professionals who are concerned with chemistry are engaged in a remarkably diversified range of activities, like monitoring the quality of a public water supply, flameproofing children' s pajamas, developing a treatment for a disease, processing a photographic film, analyzing body fluids to diagnose an illness, and determining the authenticity of a painting.

2. 大量使用被动句

科技人员在研究和解决科技问题时尊重客观规律,重视事实和方法、性能和特征、规律和推理。因此,化工英语文章大多只描述行为、活动、事实本身,并不关注行为的施动者。运用被动语态可以强调和突出所描述的对象,使文章内容更客观,减少一些主观印象。被动语态数量较多,可以占到全部谓语动词的1/3~1/2。此外,科技文章将主要信息前置,放在主语部分,从而使句子更为简短、紧凑,使行文更加准确、严谨和精练,这也是广泛使用被动态的主要原因。例如:"A bubble tower is used to distil petroleum."(蒸馏石油时使用泡罩塔)。而不用"We use a bubble tower to distil petroleum."(我们用泡罩塔蒸馏石油)。

又如:

If an oscillatory motion **was superimposed** on steady shearing, the maximum torque on the top plate during the combined motion was scarcely more than the torque in steady shearing alone; the minimum was considerably less. The average torque was much less in the combined motion and it is possible that the average rate at which work **was absorbed** was less in the combined motion than in the steady shearing alone.

这段英语是典型的科技英语,7个主语都是"非人称"的,有2个子句使用了被动语态,文体是正式的书面体,所有单词几乎全部是科技词语。

再如:

The chemical industry **is concerned** with converting raw materials, such as crude oil, firstly into chemical intermediates, and then into a tremendous variety of other chemicals. These **are then used to** produce consumer products, which make our lives more comfortable or, in some cases such as pharmaceutical products, help to maintain our well-being or even life itself. At each stage of these operations value **is added to** the product and **provided** this added value exceeds the raw material plus processing costs then a profit will **be made on** the operation. It is the aim of chemical industry to achieve this。

这段话使用了大量被动语态。

3. 大量使用名词化结构

名词化(nominalization)是指把动词、形容词通过一定方式,如加缀、转化等转换成名词的语法过程。名词化结构既可起名词的作用,也可以表达谓语动词或形容词所表达的内容,常伴有修饰成分或附加成分,构成短语。因此,语法上小到单词、词组,大到句子、段落,都可以进行名词化处理,从而形成多种组合方式的名词化短语。运用名词化结构,可以使化工英语文章行文简洁、内容确切、文体正式,从而能更好地表达现象、事实、特点以及抽象化的逻辑思维。

例如:

If the experiment is done by this method, there will be some loss of equipment.

如果用这种方法做实验,设备会有一些损失。

改成名词化结构:

The doing of the experiment by this method will entail some loss of equipment.

又如:

Chemical engineers are responsible for defining and designing the processes of which the purpose is to produce, transform and transport the materials.

化学工程师主要负责对以物料的生产、转换和运输为目的的过程进行概念界定和设计。

改成名词化结构:

Chemical engineers are responsible for the conception and design of processes for the purpose of production, transformation and transport of materials.

【练习0.1】 将下列英语翻译成汉语(特别注意加粗部分)。

(1) **The shrinkage of the sun** to this state would transform our oceans into ice and our atmosphere into liquid air.

译:

(2) **The accumulation of new data** during the past decade has brought a refinement of some earlier views and concepts.

译:

(3) **The rotation of the earth** on its own axis causes the change from day to night.

译:

(4) Television is **the transmission and reception of images** of moving objects by radio waves.

译:

(5) **The testing of machines by this method** entails some loss of power.

译:

4. 大量使用后置定语

放在被修饰词后、用来表示这个名词或代词的部分称为后置定语。化工英语中经常使用后置定语，常见的主要有4类：定语从句、非谓语动词短语、介词短语和形容词短语。例如：

(1) The proportion of the various ingredients which go into concrete, the way it is mixed, and even the water which is used are very important to the finished material.

制作混凝土所用的各种配料的比例、搅拌的方法乃至所用的水，对成品材料来说都是十分重要的。

定语从句which go into concrete... 和which is used... 做后置定语分别修饰ingredients和water。

(2) Oxidizers based on ammonium nitrate are chiefly characterized by its low signature.

硝酸铵基氧化剂的主要特点是它的信号特征低。

非谓语动词(过去分词短语)based on ammonium nitrate做后置定语修饰oxidizers。

(3) The choice of material in construction of bridges is basically between steel or concrete and main trouble with concrete is that its tensile strength is very small.

桥梁建筑材料基本上仍在钢材和混凝土之间选择，而混凝土的主要缺点是抗拉强度低。

介词短语in construction of bridges和with concrete做后置定语分别修饰material和trouble。

(4) The technologies and processes available to manufacture these products are by fractionation, hydrogenation and interesterification.

该技术和工艺可以实现这些产品的分馏、加氢和交换。

形容词available做后置定语修饰technologies和processes。

【练习0.2】 将下列英语翻译成汉语(特别注意加粗部分)。

(1) As a ship is loaded, it sinks deeper into the water, displacing an additional amount of water **equal to the weight of the added load.**

译：

(2) The forces **due to friction are** called frictional forces.

译：

(3) A call **for paper** is now being issued.

译：

(4) Iron ore is an iron oxide **not very different from ordinary iron rust.**

译：

(5) In this factory the only fuel **available** is coal.

译:

5. 大量使用非谓语动词

动词的非谓语形式容易阐明各个事物之间的关系、事物的位置和状态变化，而且能用扩展的成分对所修饰的词进行严格的说明和限定。如机器、产品、原料等的运动、来源、型号，产品的加工手段、工艺流程和仪器、设备的操作方法，这些都要求叙述严谨、准确。每一个分词短语都能代替一个从句，因此可以完整、准确地表达某一概念和事物，同时保证行文简练、结构紧凑。为此，往往使用分词短语代替定语从句或状语从句，使用分词独立结构代替状语从句或并列分句，使用不定式短语代替各种从句，使用介词+动名词短语代替定语从句或状语从句。这样既可以缩短句子，又比较醒目。例如：

(1) **To determine the optimum parameters** will be time consuming.

最佳参数的确定十分费时。

(2) **Expanding gases** can be used **to operate a machine**.

膨胀的气体可用来发动机器。

(3) **Confined in a rigid container**, gas will expand at high temperature.

即使封闭在刚性容器中，气体在高温下也会膨胀。

(4) **Annealing** is one of the heat treatments of steel.

译:

(5) **Radiating from the earth**, heat causes air currents to rise.

译:

(6) There are different ways of **changing energy from one form into another**.

译:

(7) **Emitting infrared rays** is an important way for the human body **to give out surplus heat**.

译:

材料阅读

Curriculum of Chemical Engineering

As chemical engineering knowledge developed, it was inserted into university courses and curricula. Before World War I, chemical engineering programs were distinguishable from chemistry programs in that they contained courses in engineering drawing, engineering thermodynamics, mechanics, and hydraulics taken from engineering departments. Shortly after World WarI the first text in unit operations was published. Courses in this area became the core of chemical engineering teaching.

By the mid-1930s, chemical engineering programs included courses in (1) stoichiometry (using material and energy conservation ideas to analyze chemical process steps), (2) chemical processes or "unit operations", (3) chemical engineering laboratories "in which equipment was operated and tested", (4) chemical plant design (in which cost factors were combined with technical elements to arrive at preliminary plant designs). The student was still asked to take the core chemistry courses, including general, analytical, organic and physical chemistry. However, in addition, he or she took courses in mechanical drawing, engineering mechanics, electric circuits, metallurgy and thermodynamics with other engineers.

Since World War II chemical engineering has develop rapidly. As new disciplines have proven useful, they have been added to the curriculum. Chemical engineering thermodynamics became generally formulated and taught by about 1945. By 1950, courses in applied chemical kinetics and chemical reactor design appeared. Process control appeared as an undergraduate course in about 1955, and digital computer use began to develop about 1960.

The idea that the various unit operations depended on common mechanisms of heat, mass and momentum transfer developed about 1960. Consequently, courses in transport phenomena assumed an important position as an underlying, unifying basis for chemical engineering education. New general disciplines that have emerged in the last two decades include environmental and safety engineering, biotechnology, and electronics manufacturing processing. There has been an enormous amount of development in all fields, much of it arising out of more powerful computing and applied mathematics capabilities.

1. Science and Mathematics Courses

(1) Chemistry

Chemical engineers continue to need background in organic, inorganic and physical chemistry, but also should introduced to the principles of instrumental analysis and biochemistry.

• Valuable conceptual material should be strongly emphasized in organic chemistry including that associated with biochemical process.

• Much of thermodynamic is more efficiently taught in chemical engineering, and physical chemistry should include the foundations of thermodynamic.

(2) Physics

(3) Biology

Biology has emerged from the classification stage, and modern molecular biology holds great promise for application. Future graduates will become involved with applying this knowledge at some time in their careers.

• A special course is required on the functions and characteristics of living cells with some emphasis on genetic engineering as practiced with microorganisms.

(4) Materials Science

• Course work should include the effects of microstructure on physical, chemical, optical, magnetic and electronic properties of solids.

• Fields of study should encompass ceramics, polymers, semiconductors, metals, and composites.

(5) Mathematics, Computer Instruction

• Although students should develop reasonable proficiency in programming, the main thrust should be that use of standard software including the merging of various programs to accomplish a given task. Major emphasis should be on how to analyze and solve problems with existing software including that for simulation to evaluate and check such software with thoroughness and precision.

• Students should learn how to critically evaluate programs written by others.

• All courses involving calculations should make extensive use of the computer and the latest software. Such activity should be more frequent as students progress in the curriculum. Adequate computer hardware and software must be freely available to the student through superior centralized facilities and/or individual PC's. Development of professionally written software for chemical engineering should be encouraged.

2. Chemical Engineering Courses

(1) Thermodynamics

• The important concepts of the courses should be emphasized; software should be developed to implement the concepts in treating a wide variety of complex, yet interesting, problems in a reasonable time. The value of analysis of units and dimensions in checking problems should continue to be emphasized.

• Examples in thermodynamics should involve problems from a variety of industries so that the subject comes alive and its power in decision making is clearly emphasized.

(2) Kinetics, Catalysis, and Reactor Design and Analysis

• This course also needs a broad variety of real problems, not only design but also diagnostic and economic problems. Real problems involve real compounds and the chemistry related to them.

• Existing software for algebraic and differential equation solving make simulation and design calculation on many reactor systems quite straightforward.

• Shortcut estimating methods should be emphasized in addition to computer calculations.

• The increased production of specialties make batch ad semibatch reactor more important, and scale-up of laboratory studies is an important technique in the fast-moving specialties business.

3. Unit Operations

The unit operations were conceived as an organized means for discussing the many kind of equipment-oriented physical processes required in the process industries. This approach continues to be valid. Over the years some portions have bee given separate status such as transport phenomena and separations while some equipment and related principles have not been included in the required courses, as is the case with polymer processing, an area in which all chemical engineers should have some knowledge.

• Transport phenomena principles can be made more compelling by using problems form a wide range of industries that can be analyzed and solved using the principles taught.

• Some efficiency may be gained by teaching several principles and procedures for developing specifications and selection the large number of equipment items normally purchased off-the-shelf or as standard design.

• A great deal of time can be saved in addressing designed equipped such as fractionators and absorber be emphasizing rigorous computer calculations and the simplest shortcut procedures. Most intermediate calculation procedures and graphical methods should be eliminated unless they have real conceptual value.

(1) Process Control

• This course should emphasize control strategy and precise measurement in addition to theory.

• Some hands-on experience using current practices of computer control with industrial-type consoles should be encouraged.

• Computer simulation of processes for demonstration of control principles and techniques can be most valuable, but contact with actual control devices should not be ignored.

(2) Chemical Engineering Laboratories

• Creative problem solving should be emphasized.

• Reports should be written as briefly as possible; they should contain an executive summary with clearly drawn conclusions and brief observations and explanations with graphical rather than tabular representation of data. A great deal of such graphing can be done in the laboratory on computers with modern graphics capabilities. Detailed calculations should be included in an appendix.

• Some part of the laboratory should be structured to relate to product development.

(3) Design, Economics

• In the design course in engineering, students learn the techniques of complex problem solving and decision making within a framework of economic analysis. The very nature of processes requires a system approach, the ability to analyze a total system is one of the special attributes of chemical engineers that will continue to prove most sought after in a complex technological world.

• Because of the greater diversity of interests and job opportunities, some consideration should be given to providing a variety of short design problem of greatest personal interest.

• The design approach can be most valuable in diagnosing plant problems, and some practice in this interesting area should be provided.

• Rigorous economic analysis and predictive efforts should be required in all decision processes.

• Safety and environmental considerations should also be emphasized.

• Modern simulation tools should be made available to the students.

(4) Other Engineering Courses

The electrical engineering courses should emphasize application of microprocessors, lasers, sensing devices, and control systems as well as the traditional areas of circuits and motors. The course should provide insight into the principles on which each subject is based.

Remaining courses in engineering mechanics and engineering drawing should be considered for their relevance to current and future chemical engineering practice.

4. Other Courses

(1) Economics and Business Courses

It is difficult to find a single course in economics or business departments that covers the various needs of engineers. The qualitative ability of engineers makes it possible to teach following topics in a single-semester course—in many cases in the Chemical Engineering Department: business economics, project economic analysis, economic theory, marketing and market studies, and national and global economics.

(2) Humanities and Social Science Courses

It is important to understand the origins of one's own culture as well as that of others.

(3) Communication Course

Since improved communication skills require continuous attention, the following requirements may be useful:

• Oral presentations in at least one course each year.

• Several literature surveys in the junior and senior years.

• Introduce computer-based communication systems.

(4) Area of Specialization

The elective areas should be generous in hours to maximize freedom of choice. Each department will have to consider its own and its total university resources and strengths as well as the quality and preparation of its students. The suggested areas are:

• Life sciences and applications.

• Materials sciences and applications.

• Catalysis and electrochemical science and applications.

• Separations technology.

• Computer applications technology.

• Techniques of product development and marketing.

• Polymer technology.

Each of these areas should be strongly career-oriented. The interest in a given area will depend on opportunities perceived by the students.

（摘自 胡鸣，刘霞. 化学工程与工艺专业英语[M]. 北京：化学工业出版社，2007.）

第1篇

化工专业英语词汇

第1章 构 词 法

化工专业英语术语的构成规则主要有以下几种：合成法(composition)、转化法(conversion)、派生法(derivation)、压缩法(shortening)、混成法(blending)、符号法(signs)、字母象形法(letter symbolizing)。

1.1 合 成 法

将两个及以上的单词合成为一个单词的方法叫作合成法，主要有下面几种情况：

1.1.1 合成名词

1. 名词+名词

由两个及以上的名词构成一个合成名词，前面的名词解释、说明后面的名词，单词的中心意思由后一个名词表达，它们之间表示着各种关系，例如：

copper pipe / pipe made of copper　铜管(表示由什么材料制成)

angle steel / steel whose sides are perpendicular to each other　角钢(表示材料的形状)

rust resistance / ability to resist rust　防锈(表示动作对象)

fluid mechanics / mechanics of fluid　流体力学(表示所属关系)

pollution problem / problem concerning pollution　污染问题(表示有关方面)

2. 形容词+名词

形容词+名词的修饰关系是前者修饰后者，例如：

blue print　方案　　periodic table　周期表

mixed powder　混合粉末　　atomic weight　原子量

3. 动名词+名词

动名词所表示的是与被修饰词有关的动作，而名词所表示的是可用的场所或物品，例如：

launching site　发射场

flying suit　飞行衣

navigating instrument 导航仪

4. 名词+动名词

例如：

paper-making 造纸

ship-building 造船

machine-shaping 加工成型

5. 其他构成方式

例如：

by-product 副产品(介词+名词)

make-up 化妆品(动词+副词)

out-of-door 户外(副词+介词+名词)

pick-me-up 兴奋剂(动词+代词+副词)

1.1.2 合成形容词

1. 形容词+名词

例如：

new-type 新型的

long-time 持久的，长期的

quick-change 快速变化的

2. 名词+分词

名词+分词结构主要有2种形式：

(1) 名词+现在分词：有主动含义，其中名词相当于动作的宾语，例如：time-consuming(费时间的)。

(2) 名词+过去分词：有被动含义，其中名词相当于动作的发出者，例如：arts-oriented(倾向文科的)、family-oriented(注重家庭的)。

3. 副词+分词

在副词+分词结构中，副词通常表示程度、状态。

例如：

hard-working 勤劳的

far-ranging 远程的

well-rounded 素质全面的

newly-invented 新发明的

4. 形容词+分词

例如：

Free-cutting 易切削的

direct-acting　直接作用的

ready-made　现成的

ill-equipped　装备不良的

5. 名词+形容词

此类合成形容词中的名词,有时是比喻的对象,例如:

paper-thin　薄如纸的

colour-blind　色盲的

skin-deep　肤浅的

1.2　转　化　法

在现代英语的发展过程中,基本上摒弃了词尾的变化,可以把一个单词直接由一种词类转化为另一种词类,这种构词法称为转化法。其基本特点是保持了原来的词形,但改变了原来的词性,词义基本不变或稍有引申。

1.2.1　由名词转化为动词

1. 某些抽象名词转化为动词

form　*n.* 形式⟶*v.* 形成

heat　*n.* 热⟶*v.* 加热

power　*n.* 动力⟶*v.* 用动力发动

knowledge　*n.* 知识⟶*v.* 了解,知道

2. 某些物品名称转化为动词

machine　*n.* 机器⟶*v.* 加工

bottle　*n.* 瓶子⟶*v.* 瓶装

oil　*n.* 油⟶*v.* 加油

picture　*n.* 图画⟶*v.* 描绘

1.2.2　由形容词转化为动词

clean　*adj.* 干净的⟶*v.* 使……干净,清洁

dry　*adj.* 干燥的⟶*v.* 干燥,弄干

better　*adj.* 好的⟶*v.* 改善

1.2.3 由副词转化为动词

up *adv*. 向上⟶ *v*. 提高
back *adv*. 向后⟶ *v*. 倒车,后退
forward *adv*. 向前⟶ *v*. 推进

1.2.4 由动词转化为名词

flow *v*. 流动⟶ *n*. 流量
stand *v*. 站立⟶ *n*. 支架,看台

1.3 派 生 法

派生法是通过加前、后缀构成一个新词。这是化工类科技英语中最常用的构词法。例如在有机化学中,烷烃就是用前缀(如拉丁语或希腊语前缀)表示分子中的碳原子数,再加上"-ane"作词尾构成的。若将词尾变成"-ene""-yne""-ol""-al""-one""-yl",则分别表示"烯""炔""醇""醛""酮""基"等。以此类推,从而构成千万种化学物质名词。常遇到这样的情况,许多化学化工名词在英语词典上查不到,但若掌握这种构词法,通过分析其前、后缀分别代表的意思,合在一起即是该词的意义,如表1.1中烷烃和烷基的词汇构成。

Table 1.1 The vocabulary of alkane and alkyl (C_1~C_{10})
表1.1 烷烃和烷基的词汇构成(C_1~C_{10})

数字	拉丁语或希腊语前缀	烷烃(-ane)	烷基(-yl)
1	mono-	methane	methyl
2	di-,bi-	ethane	ethyl
3	tri-	propane	propyl
4	tetra-,quadri-	butane	butyl
5	pent(a)-	pentane	pentyl
6	hex(a)-	hexane	hexyl
7	hept(a)-	heptane	heptyl
8	oct(a)-	octane	octyl
9	non(a)-	nonane	nonyl
10	dec(a)-	decane	decyl

1.3.1 前缀

前缀一般具有一定的含义，因此加前缀一般会改变词义，但不改变词性，一个词或词根可以有几个前缀。

1. 表示否定意义“不”“无”“非”等的前缀

a-：asexual 无性别的； atypical 非典型的； asymmetry 不对称的

dis-：disproportion 歧化； diversity 多样性； dissimilar 不同的

im-：immaterial 非物质的，无形的； impurity 杂质

ir-：irrevocable 不可撤销的

un-：umremarkable 不显著的，普通的； unsystematic 无系统的

in-：incoherent 不相干的； informal 非正式的； incomplete 不完全的

non-：nonuniform 不统一的

2. 表示意义相反的前缀

anti-：anticorrosion 抗腐蚀的； antistatic 抗静电的； antimicrobial 抗菌的； anticlockwise 逆时针方向的

contra-：contrary 相反的，不利的； contramissile 反弹道导弹； contradict 否认，反驳； contraposition 对立

counter-：counteract 对抗，抵消； counteraction 反作用； countermand 撤销(命令)； counterbalance 平衡； countercurrent 对立

3. 表示数字和数量的前缀

(1) 表示“半”“一半”

demi-：demicontinuous 半连续的

semi-：semi-insulating 半绝缘的； semicircle 半圆； semiconductor 半导体

hemi-：hemisphere 半球； hemiplegia 半身不遂，偏瘫

(2) 表示“单一”

uni-：unicoil 单线圈； unipolar 单极的； unidirectional 单向性的； unicellular 单细胞的

mono-：monochromatic 单色光的； monoatomic 单原子的； monoxide 一氧化物

(3) 表示“二”“双”

di-：diacid 二价酸的，二酸的； diatomic 双原子的

(4) 表示单位

kilo-(千)：kilogram 千克； kilometer 千米； kilocalorie 千卡(热量)； kiloampere 千安培

deci-(分)：decigram 分克(1/10克)； decimeter 分米

centi-(厘)：centimeter 厘米； centigram 厘克(10^{-2}克)

nano-(纳)：nanometer 纳米； nanosecond 纳秒(10^{-9}秒)
milli-(毫)：millimeter 毫米； millilitre 毫升； milligram 毫克
micro-(微)：micrometer 微米(测微计、千分尺)； microsoft 微软
poly-(多、聚)：polymer 聚合物； polyfunctional 多功能的； polyacid 多酸的

4. 表示方位的前缀

super-：supervise 临视(super+vise 看→在上面看→临视)
superstructure 上层建筑(super+structure 结构)
superimpose 放……上面，强加(super+impose 放上去→在上面放→强加)
superficial 肤浅的[super+fic(做)+ial→在表面上做]
superintend 监督(super+intend 关心→在上面关心→监督)
supersede 淘汰，取代(super+sede 坐→坐上去→淘汰以前的东西)
superstition 迷信[super+stit站+ion→站在人(理智)之上的东西→迷信]
par(a)-(对位)：paradichlorobenzene 对二氯苯
ortho-(邻位)：orthodichlorobenzene 邻二氯苯
met(a)-(间，中)：metadichlorobenzene 间二氯苯
inter-(在中间)：interatomic 原子间的
intra-(在内)：intramolecular 分子内的
over-(在上面)：overbridge 天桥； overwrite 写在……上面

5. 表示超过以前的前缀

super-：supersized 超大型的
supersonic 超音速的(super+sonic 声音的→超音速的)
supernatural 超自然的(super+natural 自然的→超自然的)
superfluous 多余的[super+flu(流)+ous→流出太多]
supercilious 目中无人的[super+cili(眉毛)+ous→在眉毛上看人→目中无人]
supersensitive 过度敏感的(super+sensitive 敏感的→过度敏感的)
supercharge 负载过重[super+charge(收费，负担)→负担过重]
over-：overemphasize 过分强调； overeat 暴饮暴食； overeducate 过度教育
ultra-：ultrasonic 超声的； ultraviolet 紫外的； ultrared(infrared) 红外的
extra-：extrahazardous 特别危险的； extramolecular 在分子以外的

6. 其他

en-(使)：enlarge 扩大； enlighten 启发，开导； enwind 缠绕
re-(更新，再)：reproduce 再生产； reprint 再版； renumber 重编号码
sub-(亚，次)：subgroup 亚层； subsurface 下表面； subtropics 亚热带
mis-(错，误)：misarrange 排错； misunderstand 误解； mistranslate 误译
auto-(自动)：automation 自动化； autotimer 自动定时器； autosuggestion 自我暗示
inter-(相互)：interaction 相互作用； interview 会见； interweave 混纺

hydro-(水,液,含氢的): hydroaeroplane 水上飞机; hydrocarbon 碳氢化合物
photo-(光): photoactive 光敏性的; photocatalysis 光催化; photochemical 光化学的
thermo-(热): thermocouple 热电偶; thermostability 热稳定性; thermoanalysis 热分析
electro-(电): electrobath 电解液; electrochemistry 电化学; electrode 电极

1.3.2 后缀

词或词根加上后缀所构成的派生词,其词义基本保持不变,但词性往往发生改变。例如:

chlorine 氯(气)
chloric 氯的,含氯的
chloride 氯化物
chlorinate 氯化,使……与氯化合/反应

1. 构成名词的后缀

(1) 表示“人”“器具”

-or: supervisor 监督人; translator 翻译家; protector 保护者
-er: computer 电脑; worker 工人; turner 车工; tinner 锡匠

(2) 表示“人”“主义者”

-ist: scientist 科学家; chemist 化学家; geologist 地质学家

(3) 表示行为

-ment: movement 运动; medicament 用药物治疗
-tion: calculation 计算; consideration 考虑; preparation 准备
-sion: transmission 传递; discussion 讨论
-age: leakage 泄露; barrage 堤,坝; garbage 垃圾,废物

(4) 表示性质、状态等概念

-th: length 长度
-ty: safety 安全
-ity: facility 能力; personality 个性; ductility 延性
-ure: departure 分别; pressure 压力
-ence: confidence 信任; difference 区别; existence 存在
-ance: resistance 阻力; ignorance 无知
-ness: blindness 盲目; weakness 弱点; usefulness 有效性
-ability: adaptability 适应性; useability 可用性; readability 可读性; malleability 延展性
-ibility: reversibility 可逆性; sensibility 敏感性; conductibility 传导性

2. 构成动词的后缀

-ize(……化)：fuctionalize 功能化； industrialize 工业化； normalize 正常化

-fy/-ify(使……成为)：simplify 简化； gasify 气化； purify 净化； electrify 电气化

-en(使……变得)：strengthen 使……坚固； soften 软化，退火； harden 硬化，淬火

3. 构成副词的后缀

-ly(……地)：carefully 仔细地； conditionally 有条件地； practically 实际上

-ward(s)(……向)：backward 向后，逆向地； upwards 朝上； downwards 向下

-wise(表示方向)：clockwise 顺时针方向地

4. 构成形容词的后缀

-y：hardy 坚固的

-al：frictional 摩擦的； periodical 周期性的

-ic：energetic 高能的

-ant：resistant 抗……的； assistant 辅助的

-ent：consistent 一致的； dependent 依赖的

-ous：continuous 连续的； poisonous 有毒的

-like：wavelike 波状的； steelike 钢铁般的

-able：changeable 可变的； inflammable 易燃的

-ible：convertible 可逆的； flexible 易弯曲的

-ful：useful 有用的； peaceful 和平的

-less：colorless 无色的； useless 无用的

1.4 压 缩 法

1. 只取单词首字母

EST: english for science and technology 科技英语

CPU: central processing unit 中央处理器

ppm：part per million 百万分之一

SHE：standard hydrogen electrode 标准氢电极(或NHE：normol hydrogen electrode)

PM：particulate matter 细颗粒物

2. 删去单词中部分字母

amp：ampere 安培

kg：kilogram 千克

km：kilometer 千米

phone：telephone 电话

quake：earthquake 地震

fridge：refrigerator 冰箱

gas：gasoline 汽油

lab：labrary 图书馆

maths: mathematics　数学　　exam: examination　考试

1.5 混　成　法

该方法是将两个词的一头一尾连在一起，构成一个新词。例如：

smog: smoke + fog　烟雾

motel: motor + hotel　汽车旅馆

positron: positive + electron　正电子

telex: teleprinter + exchange　电传

transistor: transfer + resistor　晶体管

medicare: medical + care　医疗保障

modem: modulator + demodulator　调制解调器

aldehyde: alcohol + dehydrogenation　醛（醇+脱氢）

1.6 符　号　法

&: and　和

/: and 或 or，"和"或"或"（例如：M/N——M 和 N，M 或 N）

#: number　号码（例如：# 9= No. 9= number 9）

$: dollar　美元，加元

1b: pound　英镑

￥: yuan　元（人民币）

1.7 字母象形法

字母象形法构成词的模式是：大写字母+连字符+名词，用以表示事物的外形，产品的型号、编号等。英译汉时，注意采用形译法，亦可根据具体情况翻译。例如：

I-bar / I-steel　工字铁　　I-shaped　工字形

T-square　丁字尺　　T-beam　T 字梁

V-belt　V 带　　X-ray　X 射线

n-region　n 区 p-region　p 区　　P-N-junction　P-N 结

T-connection　T字连接(三通)　U-shaped magnet　马蹄形磁铁

【练习1.1】 化学化工英语的构词方法主要有哪些?

【练习1.2】 翻译:马蹄形磁铁、醛、热电偶、电解质、甲醛、甲烷、乙酸、丙烯、丁基、抗辐射的、聚合物、聚噻吩、聚苯胺、光学的、电学的、有机的、微米、毫克、微生物、发光的、不能破坏的、无穷的、交互式的、可逆的、软管、淬火、硬化、电致发光、发射、生物技术、生物传感器、电容器、传导性、抗静电的。

材料阅读

How to Keep a Laboratory Notebook

A lab notebook is the way real scientists keep track of their work. It may seem tedious or even unnecessary to you, but it is an important part of any lab experience. The notebook should be complete enough that you could refer back to it in a few years and repeat the experiments.

General Guidelines:

(1) The notebook must be **permanently bound**: no loose-leaf or spiral notebooks.

(2) Handwriting must be legible. Your TA(teaching assistant) will not grade materials that he or she cannot easily read. All notes should be taken in pen with the exception of colored drawings that may be done with pencils. Errors should be crossed through with a single line, not erased or obliterated.

(3) All information in the notebook must be handwritten or represent actual results, such as photographs. Do not place any photocopied material into your notebook unless specifically directed to do so.

(4) Everything you do in the laboratory should be recorded in your lab notebooks, including notes, drawings, data, speculations, etc. Everything from your initial strategy through planning, execution and interpretation and should be in your notebook.

(5) Keep all of your lab-related notes, including lab lecture notes, in one notebook. Keep a separate binder for the lab manual and lab handouts.

(6) Keep in mind that reports and presentations will be prepared from the notebook. You should have much more information recorded in your notebook than you can or should put on a poster or into a presentation.

The notebook should include:

(1) The first two pages reserved for a table of contents.

(2) Notes from lab lectures, discussions and your own research.

(3) Answers to assigned questions.

(4) Prelab Section for experiments:

• **Title** of experiment and **date**.

• The **Objective(s)** of the lab: what you are trying to do and why you are trying to do it.

• The **Procedure** in flow chart or outline form. This should not be an exact copy of the lab manual instructions, but reworked in a manner easy for you to follow.

(5) Any **deviations** from your written procedure. This includes changes both intentional and accidental.

(6) **Observations:** everything that happens during your experiment that may have a bearing on the outcome or interpretation of the experiment (this includes color, precipitate, time, temperature, etc).

(7) **Data:** raw and calculated. Use complete sentences, tables and graphs where appropriate. Show sample calculations with steps and units.

(8) **Discussion:** Interpret your results. Refer back to your predictions. Draw **conclusions** about experiment. Make suggestions for further experiments or refinements to the procedure.

（摘自 http://www.colorado.edu/MCDB/MCDB3140/notebooks.html.）

第2章　无机化学的词汇构成

Inorganic Chemistry

Inorganic chemistry is the branch of chemistry concerned with the properties and reactions of inorganic compounds. This includes all chemical compounds except the many which are based upon chains or rings of carbon atoms, which are termed organic compounds and are studied under the separate heading of organic chemistry. The distinction between the two disciplines is not absolute and there is much overlap, most importantly in the sub-discipline of organometallic chemistry.

2.1　元素和单质的命名

“元素”和“单质”的英文翻译都是“element”，有时为了区别，在强调“单质”时可用“free element”。因此，单质的英文名称与元素的英文名称是一样的。

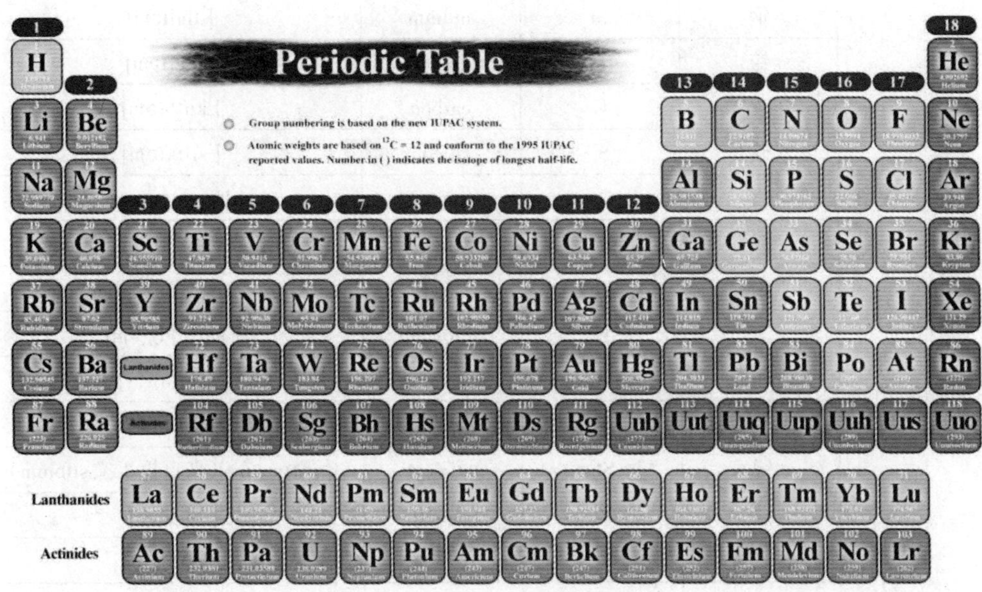

Figure 2.1　Periodic table of elements

图2.1　元素周期表

图2.1中的元素周期表根据原子序数从小至大排序，图中一横行为一个周期，一列为一族，共有7个周期，16个族。根据价电子所在轨道可以分为5个区：s区、p区、d区、ds区、f区。以下主要讨论s区和p区元素（见表2.1）。

Table 2.1 Elements and their English names

表2.1 元素及其英文全称

区	族	元素	元素符号	英文全称	英标
s	IA	氢	H	hydrogen	[ˈhaɪdrədʒ(ə)n]
		锂	Li	lithium	[ˈlɪθɪəm]
		钠	Na	sodium	[ˈsəʊdɪəm]（来自拉丁文natrium）
		钾	K	potassium	[pəˈtæsɪəm]（来自拉丁文kalium）
		铷	Rb	rubidium	[rʊˈbɪdɪəm]
		铯	Cs	cesium	[ˈsiːzɪəm]
		钫	Fr	francium	[ˈfrænsɪəm]
	IIA	铍	Be	beryllium	[bəˈrɪlɪəm]
		镁	Mg	magnesium	[mægˈniːzɪəm]
		钙	Ca	calcium	[ˈkælsɪəm]
		锶	Sr	strontium	[ˈstrɒntɪəm; ˈstrɒnʃ(ɪ)əm]
		钡	Ba	barium	[ˈbeərɪəm]
		镭	Ra	radium	[ˈreɪdɪəm]
p	IIIA	硼	B	boron	[ˈbɔːrɒn]
		铝	Al	aluminium	[æl(j)ʊˈmɪnɪəm]
		镓	Ga	gallium	[ˈgælɪəm]
		铟	In	indium	[ˈɪndɪəm]
		铊	Tl	thallium	[ˈθælɪəm]
	IVA	碳	C	carbon	[ˈkɑːb(ə)n]
		硅	Si	silicon	[ˈsɪlɪk(ə)n]
		锗	Ge	germanium	[dʒɜːˈmeɪnɪəm]
		锡	Sn	tin	[tɪn]（来自拉丁文stannum）
		铅	Pb	lead	[liːd]（来自拉丁文plumbum）
	VA	氮	N	nitrogen	[ˈnaɪtrədʒ(ə)n]
		磷	P	phosphorus	[ˈfɒsf(ə)rəs]
		砷	As	arsenic	[ˈɑːs(ə)nɪk]
		锑	Sb	antimony	[ˈæntɪmounɪ]（来自拉丁文stibium）
		铋	Bi	bismuth	[ˈbɪzməθ]
	VIA	氧	o	oxygen	[ˈɒksɪdʒ(ə)n]
		硫	s	sulfur	[ˈsʌlfə]

续表

区	族	元素	元素符号	英文全称	英标
p	VIA	硒	Se	selenium	[sɪˈliːnɪəm]
		碲	Te	tellurium	[teˈljʊərɪəm]
		钋	Po	polonium	[pəˈlonɪəm]
	VIIA	氟	F	fluorine	[ˈflɔrin]
		氯	Cl	chlorine	[ˈklɔrin]
		溴	Br	bromine	[ˈbrəʊmiːn]
		碘	I	iodine	[ˈaɪədiːn]
		砹	At	astatine	[ˈæstətiːn]
	0	氦	He	helium	[ˈhiːlɪəm]
		氖	Ne	neon	[ˈniːɒn]
		氩	Ar	argon	[ˈɑːgɒn]
		氪	Kr	krypton	[ˈkrɪptɒn]
		氙	Xe	xenon	[ˈzenɒn]
		氡	Rn	radon	[ˈreɪdɒn]
过渡元素（部分）		铁	Fe	iron	[ˈaɪən]（来自拉丁文 ferrum）
		锰	Mn	manganese	[ˈmæŋgəniːz]
		铜	Cu	copper	[ˈkɒpə]（来自拉丁文 cuprum）
		锌	Zn	zinc	[zɪŋk]
		汞	Hg	mercury	[ˈmːkjəri]（来自拉丁文 hydrargyrum）
		银	Ag	silver	[ˈsɪlvə]（来自拉丁文 argentums）
		金	Au	gold	[gəʊld]（来自拉丁文 aurum）

2.2　阳离子的命名

2.2.1　单一价位阳离子

单一价位阳离子的名命原则为

元素名称 + 离子（element + ion）

IA、IIA 元素和 Al 的离子具有固定的化合价，这些元素的离子命名在元素后加“ion”。例如：

Na^+:sodium ion

H^+:hydrogen ion

Li^+:lithium ion

K^{+}:potassium ion

Rb^{+}:rubidium ion

Be^{2+}:beryllium ion

Mg^{2+}:magnesium ion

Ca^{2+}:calcium ion

Ba^{2+}:barium ion

Al^{3+}:aluminum ion

2.2.2 多价位阳离子

1. 系统命名

多价位阳离子的系统命名原则为

元素(化合价罗马数字)+离子[element (N) + ion]

2. 俗称

在多价位阳离子的俗称命名原则中,高价态的以“**-ic**”结尾,低价态的以“**-ous**”结尾。例如:

Fe^{2+}:iron(Ⅱ) ion (俗名:ferr**ous** ion)

Fe^{3+}:iron(Ⅲ) ion (俗名:ferr**ic** ion)

Cu^{+}:copper(I) ion (俗名:cupr**ous** ion)

Cu^{2+}:copper(Ⅱ) ion (俗名:cupr**ic** ion)

Sn^{2+}:tin(Ⅱ) ion

Sn^{4+}:tin(Ⅳ) ion

【练习2.1】 写出下列阳离子的英语名称。

Be^{2+}、Cs^{+}、Cr^{2+}、Cr^{3+}、Cr^{6+}、Co^{2+}、Co^{3+}、Ga^{3+}、Ag^{+}、Au^{2+}、Zn^{2+}。

2.2.3 多原子阳离子

多原子阳离子的命名原则为

原子团 + 离子(radical + ion)

例如:

NH^{4+}:ammonium ion

H_3O^{+}:hydronium ion

NO_2^{+}:nitronium ion

2.3　阴离子的命名

2.3.1　单原子阴离子

单原子阴离子的命名原则为

元素名称的词干-**ide** + 离子（element's root-**ide** + ion）

例如：

Cl^-：chlor**ide** [ˈklɔːraɪd] ion

O^{2-}：ox**ide** ion

Br^-：brom**ide** [ˈbrəʊmaɪd] ion

I^-：iod**ide** [ˈaɪədaɪd] ion

S^{2-}：sulf**ide** [ˈsʌlfaɪd] ion

H^-：hydr**ide** [ˈhaɪdraɪd] ion

氰根（CN^-）和氢氧根（OH^-）可视为单原子阴离子。例如：

CN^-：cyan**ide** [ˈsaɪənaɪd] ion

OH^-：hydrox**ide** [haɪˈdraksaɪd] ion

2.3.2　酸根离子

如果某元素能形成一种以上的含氧阴离子，则按以下规则命名：

1. 高酸根离子

高酸根离子的命名原则为

per-中心原子的词根-**ate** + 离子（**per**-central element's root-**ate** + ion）

例如：

ClO_4^-:**per**chlor**ate** [pəˈklɔːreɪt] ion

IO_4^-:**per**iod**ate** [pɜːˈraɪədeɪt] ion

MnO_4^-:**per**mangan**ate** [pəˈmæŋgənət] ion

2. 正酸根离子

正酸根离子的命名原则为

中心原子的词根-**ate** + 离子（central element's root-**ate** + ion）

例如：

ClO_3^-:chlor**ate** [ˈklɔːreɪt] ion

IO_3^-:iod**ate** [ˈaɪədeɪt] ion

PO_4^{3-}:phosph**ate** [ˈfɒsfeɪt] ion

NO_3^-:**nitrate** [ˈnaɪtreɪt] ion

SO_4^{2-}:**sulfate** [ˈsʌlfeɪt] ion

CO_3^{2-}:**carbonate** [ˈkɑːbəneɪt] ion

3. 亚酸根离子

亚酸根离子的命名原则为

中心原子的词根-**ite** + 离子(central element's root-**ite** + ion)

例如:

ClO_2^-: **chlorite** [ˈklɔːraɪt] ion

IO_2^-: **iodite** [ˈaɪədaɪt] ion

PO_3^{3-}: **phosphite** [ˈfɒsfaɪt] ion

NO_2^-: **nitrite** [ˈnaɪtraɪt] ion

SO_3^{2-}: **sulfite** [ˈsʌlfaɪt] ion

4. 次酸根离子

次酸根离子的命名原则为

hypo-中心原子的词根-**ite** + 离子(**hypo**-central element's root-**ite** + 离子)

例如:

ClO^-: **hypo**chlorite [ˌhaɪpəʊˈklɔːraɪt] ion

IO^-: **hypo**iodite [haɪpəˈaɪədaɪt] ion

PO_2^{3-}: **hypo**phosphite [ˌhaɪpəˈfɒsfaɪt] ion

5. 偏酸根(原酸失水为“偏”酸)离子

偏酸根离子的命名原则为

meta-中心原子的词根-**ate** + 离子(**meta**-central element's root-**ite** + 离子)

例如:

PO_3^-: **meta**phosphate [ˌmetəˈfɒsfeɪt] ion

SiO_3^{2-}: metasilic**ate** ion

6. 焦酸根离子

焦酸根离子的命名原则为

pyro-中心原子的词根-**ate** + 离子(**pyro**-central element's root-**ite** + ion)

例如:

$P_2O_7^{4-}$: **pyro**phosphate [paɪrəʊˈfɒsfeɪt] ion

$S_2O_7^{2-}$: **pyro**sulfate ion

其他的前缀还有“**ortho-**”(正)、“**thio-**”(硫代)(例如,$S_2O_3^{2-}$ 翻译为**thio**sulfate ion)。

7. 含氢酸根离子

含氢酸根离子的命名原则为

氢的个数前缀-hydrogen(氢) + 去掉氢的阴离子名称

(*n*-hydrogen + anion for removing hydrogen)

例如：

HCO_3^-:hydrogen carbon**ate** ion

$H_2PO_4^-$:**bi**hydrogen phosph**ate** ion

习惯命名法用“**bi-**”前缀来表示，例如：

HSO_4^-:**bi**sulfate ion

HCO_3^-:**bi**carbonate ion

【练习2.2】 写出下列酸根离子的英语名称。

(1) $Cr_2O_7^{2-}$;(2) CrO_4^{2-};(3) BrO_2^-;(4) BrO_3^-;(5) BrO^-;(6) $H_2PO_4^{2-}$;(7) HPO_4^-。

2.4　化合物的命名

化合物的中文读法是根据化学式从右往左读，而其英语命名顺序与中文读法相反，是根据化学式**从左往右**读。表示原子个数时使用前缀“mono-”“di-”“tri-”“tetra-”“penta-”“hexa-”“hepta-”“octa-”“nona-”“deca-”，但是在不会引起歧义时，这些前缀都尽可能省去。

2.4.1　分子化合物

分子化合物(molecular compounds)总的命名规则：正价元素名称 + 负价元素名称的词干-ide，分子中各元素原子的个数用**希腊数字前缀**来表示。例如：

SF_6:sulfur **hexa**fluor**ide**

1. 金属氧化物

金属氧化物(metal oxide)的命名原则为

金属阳离子+ 氧化物(cation + oxide)

例如：

FeO:iron(II) oxide(俗称:ferrous oxide)

Fe_2O_3:iron(III) oxide(俗称:ferric oxide)

Fe_3O_4:ferroferric oxide

Pb_3O_4:**tri**lead **tetr**oxide [te'trɒksaɪd]

Na_2O_2:sodium **per**oxide [pə'rɒksaɪd]

注：非最低价的二元化合物还要加前缀，如O_2^{2-}为**per**oxide，O_2^-为**super**oxide

2. 非金属氧化物

非金属氧化物(nonmetal oxide)的命名原则为

非金属元素个数前缀 - 非金属元素+ 氧个数前缀-oxide

(*n*-nonmetal element + *n*-oxide)

例如：

CO：carbon **mono**xide

CO_2：carbon **di**oxide

SO_3:sulfur **tri**oxide [traɪˈɒksaɪd]

N_2O_3:**di**nitrogen **tri**oxide [traiˈɔksaid]

P_2O_5:**di**phosphorus **pent**oxide [penˈtɒksaɪd]

N_2O_4:**di**nitrogen **tetr**oxide [teˈtrɒksaɪd]

其中tetra-、mono-前缀中的“a”“o”在后时，“o”会省去。

注：有些物质常用俗称，如NO为nitric oxide，N_2O为nitrous oxide。

3. 非金属氢化物

除了水和氨气使用俗称water和ammonia以外，其他的非金属氢化物（nonmetal hydride）都用系统名称，命名规则根据化学式的写法不同而有所不同。

（1）对于卤族和氧族氢化物，“H”在化学式中写在前面，因此将其看成与另一元素的二元化合物。例如：

HF：hydrogen fluo**ride**

HCl：hydrogen chlo**ride**

HBr：hydrogen brom**ide**

HI：hydrogen iod**ide**

H_2S：hydrogen sulf**ide**

H_2Se：hydrogen selen**ide** [ˈselin(a)id]

H_2Te：hydrogen tellu**ride** [ˈteljʊraɪd]

（2）对于其他族的非金属氢化物，“H”在化学式中写在后面，可加后缀“**-ane**”，氮族还可加“**-ine**”。例如：

PH_3：phosph**ine** [ˈfɒsfiːn]（俗称：phosphane）

AsH_3：ars**ine** [ˈɑːsiːn]（俗称：arsane）

SbH_3：stib**ine** [ˈstibiːn，-bin]（俗称：stibane）

BiH_3：bismuth**ane** [ˈbɪzməθeɪn]

SiH_4：sil**ane** [ˈsaɪleɪn]

B_2H_6：**di**bor**ane** [daiˈbɔːrein]

【练习2.3】 写出下列分子的英语名称。

（1）MgO；（2）K_2O_2；（3）NO；（4）CO；（5）CO_2；（6）HCl；（7）FeO；（8）Fe_2O_3；（9）Fe_3O_4。

2.4.2 酸

1. 无氧酸

无氧酸（即阴离子以“**-ide**”结尾的酸）的命名原则为

hydro-词根-ic + acid

例如：

HCl：**hydro**chlor**ic** acid

H_2S：**hydro**sulfur**ic** acid

注：2.4.1小节中涉及HCl等氢化物为分子的英语构成，此为无氧酸。

【练习2.4】 写出下列分子的英语名称。

(1) HF；(2) HBr；(3) HI。

2. 含氧酸

采用前、后缀的不同组合表示不同价态的含氧酸和含氧酸根阴离子，价态相同的含氧酸及含氧酸根阴离子具有相同的前缀、不同的后缀。如果某元素能形成一种以上的含氧酸，则按以下规则命名：

(1) 高(过)酸：**per-**酸根离子中非氧元素名称的词干**-ic** acid。

(2) 正酸：酸根离子中非氧元素名称的词干**-ic** acid。

(3) 亚酸：酸根离子中非氧元素名称的词干**-ous** acid。

(4) 次酸：**hypo-**酸根离子中非氧元素名称的词干**-ous** acid(从a到d含氧原子数依次递减)。

(5) 偏酸(原酸失水为"偏"酸)：**meta-** 酸根离子中非氧元素名称的词干**-ic** acid。

(6) 焦酸(两个酸分子失水为"焦")：**pyro-** 酸根离子中非氧元素名称的词干**-ic** acid。

例如：

$HClO_4$：**per**chlor**ic** acid(ClO_4^-：**per**chlor**ate** ion)

$HClO_3$：chlor**ic** acid(ClO_3^-：chlor**ate** ion)

$HClO_2$：chlor**ous** acid(ClO_2^-：chlor**ite** ion)

HClO：**hypo**chlor**ous** acid(ClO^-：**hypo**chlor**ite** ion)

H_2SO_4：sulfur**ic** acid

H_2SO_3：sulfur**ous** acid

HNO_3：nitr**ic** acid

HNO_2：nitr**ous** acid

HPO_3：**meta**phosphor**ic** acid

注：原酸全部失水为"酐"(anhydride)。如亚硫酸酐SO_2为sulfurous acid anhydride［或sulfur dioxide(二氧化硫)］。

【练习2.5】 写出下列酸的英语名称。

(1) 碳酸；(2) 高锰酸；(3) 磷酸；(4) 亚磷酸；(5) 重铬酸；(6) 铬酸；(7) 偏硫酸；(8) 焦硫酸。

2.4.3 碱

碱(bases)的命名原则为

金属阳离子 + 氢氧根(metal cation + hydroxide)

例如:

$Al(OH)_3$:aluminum hydroxide

NaOH:sodium hydroxide

$Ca(OH)_2$:calcium hydroxide

$Ba(OH)_2$:barium hydroxide

$Co(OH)_2$:cobalt(II)hydroxide

【练习2.6】 写出下列碱的英语名称。

(1) 氢氧化铵;(2) 氢氧化钾;(3) 氢氧化镁;(4) 氢氧化铁;(5) 氢氧化亚铁。

2.4.4 盐

盐(salts)的命名原则包括不带“ion”的阳离子和阴离子名称。

1. 正盐

正盐(normal salt)命名是根据化学式从左往右分别读出阳离子和阴离子的名称。其形式为

阳离子 + 阴离子(cation + anion)

例如:

$HgSO_4$:mercury(II) sulfate

Hg_2SO_4:mercury(I) sulfate

KNO_3:potassium nitrate

Na_2CO_3:sodium carbonate

NaClO:sodium hypochlorite

$FeSO_4$:iron(II) sulfate

$KMnO_4$:potassium permanganate

CuCl:copper(I) chloride

$CuCl_2$:copper(II) chloride

2. 酸式盐

酸式盐(acidic salts)同正盐的命名原则,酸根中的“H”读作hydrogen,氢原子的个数用前缀表示。命名形式如下:

阳离子 + *n*-hydrogen + 阴离子(cation + *n*-hydrogen + anion)

例如:

$NaHSO_4$:sodium hydrogen sulfate

Na_2HPO_4:**di**sodium hydrogen phosphate

NaH_2PO_4:sodium **di**hydrogen phosphate

$NaHCO_3$:sodium hydrogen carbonate

习惯命名也常用“**bi**-阴离子”表示酸式盐,例如:

$Ca(HSO_4)_2$:calcium **bi**sulfate

$NaHCO_3$:sodium **bi**carbonate

3. 碱式盐

碱式盐(basic salts)的命名原则为

阳离子+氢氧根-阴离子(cation + hydroxy-anion)

例如:

$Cu_2(OH)_2CO_3$:dicopper(II) dihydroxycarbonate

$Ca(OH)Cl$:calcium hydroxychloride

$Mg(OH)PO_4$:magnesium hydroxyphosphate

4. 复盐

复盐(mixed salts)的命名原则同正盐,并且阳离子按英文名称的第一个字母顺序命名。其形式如下:

阳离子 + 阳离子 + 阴离子(cation + cation + anion)

例如:

$NaKSO_3$:sodium potassium sulfite

$CaNH_4PO_4$:calcium ammonium phosphate

$AgLiCO_3$:silver lithium carbonate

$NaNH_4SO_4$:sodium ammonium sulfate

$KNaCO_3$::potassuim sodium carbonate

$NaNH_4HPO_4$:sodium ammonium hydrogenphosphate

5. 水合盐

水合盐的结晶水称为water或hydrate,其命名原则为

非水化合物名称 + 数字前缀-hydrate(compounds +*n*-hydrate)

例如:

$AlCl_3\cdot 6\ H_2O$:aluminum chloride 6-water (或 aluminum chloride hexahydrate)

$AlK(SO_4)_2\cdot 12\ H_2O$:aluminium potassium sulphate 12-water

$CuSO_4\cdot 5H_2O$:copper (II) sulfate pentahydrate

2.4.5　俗称

H_2O:water

NH_3:ammonia

$CaCO_3$:marble, chalk, limestone

$Ca(OH)_2$:slaked lime

$CO_2(s)$: dry ice

NaCl: table salt

N_2O: laughing gas

CaO: quick lime

$NaHCO_3$: baking soda

$MgSO_4 \cdot 7H_2O$: epsom salt

$CaSO_4 \cdot 2H_2O$: gypsum

$Na_2CO_3 \cdot 10H_2O$: washing soda

2.5 配位化合物的命名

配位化合物简称配合物，也叫络合物(complex compound)。其在结构上由中心原子和配位体完全或部分以配位键结合而成。

2.5.1 配位体

1. 阴离子配体

阴离子配体(negative ions as ligands)的命名原则为

配体=原子的词根-o(ligand = element's root-o)

例如：

CN^-: cyano ['saɪəno]

CO_3^{2-}: carbonato

F^-: fluoro ['flʊərə]

Br^-: bromo ['brəuməu]

NO^{3-}: nitrato

NO^{2-}: nitro ['naɪtrəʊ]

O^{2-}: oxo ['ɔksəu]

Cl^-: chloro

OH^-: hydroxo

H^-: hydrido

2. 中性分子配体

中性分子配体(neutral molecules as ligand)的命名原则为

配体=原子团名称(ligand = radical name)

例如：

NH_3: ammine ['æmi:n]

CO: carbonyl ['kɑ:bənaɪl; -nɪl]

H_2O: aqua ['ækwə]

CH_3NH_2: methylamine [mi:'θaɪləmi:n]

$H_2NCH_2CH_2NH_2$: ethylenediamine [ˌɛθɪlinˌdaɪə'mɪn]

2.5.2 配位离子

1. 配位分子或带正电荷的配离子

配位分子(neutral complex)或带正电荷的配离子(complex ions with positive charge)命名

原则为

配离子=配体个数前缀-配体-金属离子(电荷数)

[complex ion = *n*-ligand-metal ion (*N*)]

例如:

$Ag(NH_3)_2^+$: **di**-ammine-silver (I)

$Cu(NH_3)_4^{2+}$: **tetra**-ammine-copper (II)

$[Co(NH_3)_3(NO_2)_3]$: **tri**-ammine-**tri**-nitro-cobalt (III)

$[Fe(CO)_5]$: **penta**-carbonyl-iron (0)

2. 带负电荷的配离子

带负电荷的配离子(complex ions with negative charge)命名原则为

配离子=配体个数前缀-配体-金属原子的词根-ate(电荷数)

[complex ion = *n*-ligand-metal's root-ate (*n*)]

例如:

$[Fe(CN)_6]^{3-}$: **hexa**-fluoro-ferrate (III)

$[AlF_8]^{3-}$: **hexa**-fluoro-aluminate (III)

$[AuCl_4]^-$: **tetra**-chloro-aurate (III)

2.5.3　配合物的命名

配合物的命名原则为

配合物=阳离子+阴离子(complex = cation + anion)

例如:

$Li[AlH_4]$:lithium **tetra**-hydro-aluminate (III)

$[Ag(NH_3)_2]Cl$:di-ammine-silver (I) chloride

$K_3[Fe(CN)_6]$:potassium **hexa**-cyano-ferrate (III)

$[Cu(NH_3)_4]SO_4$:**tetra**-ammine-copper (II) sulfate

$Ni(CO)_4$:**tetra**-carbonyl-nickel (0)

【练习2.7】 写出下列物质的英语名称。

CO	H_2SO_3	$Ca(HSO_3)_2$
CO_2	H_3PO_4	Hg_2SO_4
SO_3	$HClO$	$HgSO_4$
N_2O_3	$HClO_2$	$NaClO$
P_2O_5	$HClO_3$	$Bi(OH)_2NO_3$
Cl_2O_7	$HClO_4$	$NaKSO_4$
FeO	Na_2HPO_4	$Sn(OH)_2$
Fe_2O_3	NaH_2PO_4	$NaCl$
H_2SO_4	$NaHCO_3$	$CuCl_2$

$AlK(SO_4)_2 \cdot 12H_2O$	$SnCl_2$	$KNaCO_3$
CaO	SiH_4	$AgLiCO_3$
$AgNO_3$	B_2H_6	$[Ag(NH_3)_2]Cl$
$MgCl_2$	KBH_4	$K_4[Fe(CN)_6]$
$Al(OH)_3$	H_2S	

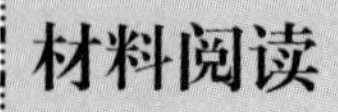

Inorganic Chemistry

The bulk of inorganic compounds occur as salts, the combination of cations and anions joined by ionic bonding. Examples of cations are sodium Na^+, and magnesium Mg^{2+} and examples of anions are oxide O^{2-} and chloride Cl^-. As salts are neutrally charged, these ions form compounds such as sodium oxide Na_2O or magnesium chloride $MgCl_2$. The ions are described by their oxidation state and their ease of formation can be inferred from the ionization potential (for cations) or from the electron affinity (anions) of the parent elements.

Important classes of inorganic compounds are the oxides, the carbonates, the sulfates and the halides. Many inorganic compounds are characterized by high melting points. Inorganic salts typically are poor conductors in the solid state. Another important feature is their solubility in e. g. water, and ease of crystallization. Where some salts (e. g. NaCl) are very soluble in water, others (e.g. SiO_2) are not.

The simplest inorganic reaction is double displacement when in mixing of two salts the ions are swapped without a change in oxidation state. In redox reactions one reactant, the oxidant, lowers its oxidation state and another reactant, the reductant, has its oxidation state increased. The net result is an exchange of electrons. Electron exchange can occur indirectly as well, e.g. in batteries, a key concept in electrochemistry.

When one reactant contains hydrogen atoms, a reaction can take place by exchanging protons in acid-base chemistry. In a more general definition, an acid can be any chemical species capable of binding to electron pairs is called a Lewis acid; conversely any molecule that tends to donate an electron pair is referred to as a Lewis base. As a refinement of acid-base interactions, the HSAB theory takes into account polarizability and size of ions.

Inorganic compounds are found in nature as minerals. Soil may contain iron sulfide as pyrite or calcium sulfate as gypsum. Inorganic compounds are also found multitasking as biomolecules: as electrolytes (sodium chloride), in energy storage (ATP) or in construction (the polyphosphate backbone in DNA).

The first important man-made inorganic compound was ammonium nitrite for soil fertilization through the Haber process. Inorganic compounds are synthesized for use as catalysts such as vanadium(V) oxide and titanium(III) chloride, or as reagents in organic chemistry such as lithium aluminium hydride.

Subdivisions of inorganic chemistry are organometallic chemistry, cluster chemistry and bioinorganic chemistry. These fields are active areas of research in inorganic chemistry, aimed toward new catalysts, superconductors, and therapies.

(摘自 http://en.wikipedia.org/wiki/Inorganic_chemistry.)

无机化学常用词汇

acetic acid 醋酸
acid 酸
acid-base indicator 酸-碱指示剂
acidic 酸性的
addition 加成
alkali 碱
alkaline 碱性的
alkaline metal 碱金属
aluminum 铝
ammonia 氨,氨水
ammonium nitrite 亚硝酸铵
anion(negative ion) 阴离子
atom 原子
atomic structure 原子结构
barium 钡
bond 键
bond angle 键角
bromine 溴
buffer 缓冲液
calcium 钙
calcium carbonate 碳酸钙
calcium phosphate 磷酸钙
calcium sulphate 硫酸钙
carbon 碳
carbon dioxide 二氧化碳
carbon monoxide 一氧化碳
inorganic compound 无机化合物
iodine 碘
carbon-carbon bond 碳-碳键
carbonic acid 碳酸
cation(positive ion) 阳离子
chemical bond 化学键
chlorine 氯
chromium 铬
coordinate bond 配位键
copper 铜
covalent bond 共价键
dissociation constant 解离常数
double bond 双键
EDTA 乙二胺四乙酸
electrolysis 电解
electron 电子
excited state 激发态
ferroin 邻二氮菲亚铁
fluorine 氟
ground state 基态
halogenation 卤化
hydrochloric acid 盐酸
hydrogen 氢
hydrogen bond 氢键
hydrogen peroxide 过氧化氢
hydrogen sulphide 硫化氢
hydrolyze 水解
inorganic chemistry 无机化学
ion 离子
ionic bond 离子键

iron 铁
lone pair 孤对电子
magnesium 镁
magnesium hydroxide 氢氧化镁
manganese 锰
metallic bond 金属键
molecular formula 分子式
nitrogen 氮
nitrous acid 亚硝酸
oxidation 氧化
oxidizing agent 氧化剂
oxygen 氧
perchloric acid 高氯酸
phenolphthalein 酚酞
phosphoric acid 磷酸
phosphorus 磷
platinum 铂
polar bond 极性键
potassium 钾
potassium bicarbonate 碳酸氢钾
potassium bromide 溴化钾
potassium hydroxide 氢氧化钾
potassium nitrate 硝酸钾
proton 质子
proton acceptor 质子接受体
proton donor 质子给体
reducing agent 还原剂
reduction 还原
silicon 硅
silver 银
silver bromide 溴化银
silver chloride 氯化银
silver nitrate 硝酸银
molecular orbital 分子轨道
molecule 分子
neutral 中性的
neutralization 中和
neutron 中子
nitration 硝化(作用)
nitric acid 硝酸
single bond 单键
sodium 钠
sodium bicarbonate 碳酸氢钠
sodium carbonate 碳酸钠
sodium chloride 氯化钠
sodium nitrite 亚硝酸钠
sodium peroxide 过氧化钠
sodium sulfite 亚硫酸钠
sodium sulphate 硫酸钠
solute 溶质
solution 溶液
solvent 溶剂
strong acid 强酸
strong base 强碱
substitution 取代
sulfur 硫
sulfuric acid 硫酸
sulphonation 磺化(作用)
synthesize 合成
the periodic table 元素周期表
valence 化合价
van der waals bond 范德华键
weak acid 弱酸
weak base 弱碱
zinc 锌
zinc oxide 氧化锌

第3章　有机化合物的命名

Organic Chemistry

The nature of organic chemistry has changed greatly since 1828. Before that time the scientific philosophy known as "Vitalism" maintained that organic chemistry was the chemistry of living systems. It maintained that organic compounds could only be produced within living matter while inorganic compounds were synthesized from non-living matter. Even the word "organic" comes from the same root as the word "organism" or "organ". However people like professor Wohler beginning in 1828 determined that it was indeed possible to synthesize organic compounds from those compounds that were considered inorganic. One of the first organic compounds synthesized from basically inorganic compounds was the compound urea which is a metabolic product of urine. It was synthesized from ammonium cyanate considered a compound produced outside of living matter and therefore considered inorganic. Since then many millions of organic compounds have been synthesized "in vitro" in other words outside living tissue.

3.1 数字前缀

有机化合物中所含的1~10个的碳原子数用甲、乙、丙、丁、戊、己、庚、辛、壬、癸表示，其英文前缀依次为 meth-、eth-、prop-、buta-、penta-、hexa-、hepta-、octa-、nona-、deca-。自11起用中文数字表示，其英文前缀见表3.1。

Table 3.1　Number prefox

表3.1　数字前缀

中文名	英文前缀	中文名	英文前缀	中文名	英文前缀
十一	undeca-, endeca-	二十六	hexacosa-	四十一	hentetraconta-
十二	dodeca-	二十七	heptacosa-	四十二	dotetraconta-
十三	trideca-	二十八	octacosa-	四十三	tritetraconta-
十四	tetradeca-	二十九	nonacosa-	四十四	tetratetraconta-
十五	pentadeca-	三十	triaconta-	四十五	pentatetraconta-

续表

中文名	英文前缀	中文名	英文前缀	中文名	英文前缀
十六	hexadeca-	三十一	hentriaconta-	四十六	hexatetraconta-
十七	heptadeca-	三十二	dotriaconta-	四十七	heptatetraconta-
十八	octadeca-	三十三	tritriaconta-	四十八	octatetraconta-
十九	nonadeca-	三十四	tetratriaconta-	四十九	nonatetraconta-
二十	eicosa-	三十五	pentatriaconta-	五十	pentaconta-
二十一	heneicosa-	三十六	hexatriaconta-	六十	hexaconta-
二十二	docosa-	三十七	heptatriaconta-	七十	heptaconta-
二十三	tricosa-	三十八	octatriaconta-	八十	octaconta-
二十四	tetracosa-	三十九	nonatriaconta-	九十	nonaconta-
二十五	pentacosa-	四十	tetraconta-	一百	hecta-

3.2 烷烃的命名

3.2.1 直链烷烃的命名

直链烷烃(alkane)的命名原则为数字前缀-**ane**(number prefix-**ane**)。表3.2列出了一些烷烃的中文名称及其对应的英文名称。

Table 3.2 The English name of alkane

表3.2 烷烃英文名称

烷烃	英文名	烷烃	英文名	烷烃	英文名
甲烷	methane	二十烷	eicosane	三十九烷	nonatriacontane
乙烷	ethane	二十一烷	heneicosane	四十烷	tetracontane
丙烷	propane	二十二烷	docosane	四十一烷	hentetracontane
丁烷	butane	二十三烷	tricosane	四十二烷	dotetracontane
戊烷	pentane	二十四烷	tetracosane	四十三烷	tritetracontane
己烷	hexane	二十五烷	pentacosane	四十四烷	tetratetracontane
庚烷	heptane	二十六烷	hexacosane	四十五烷	pentatetracontane
辛烷	octane	二十七烷	heptacosane	四十六烷	hexatetracontane
壬烷	nonane	二十八烷	octacosane	四十七烷	heptatetracontane
癸烷	decane	二十九烷	nonacosane	四十八烷	octatetracontane
十一烷	undecane	三十烷	triacontane	四十九烷	nonatetracontane
十二烷	dodecane	三十一烷	hentriacontane	五十烷	pentacontane

续表

烷烃	英文名	烷烃	英文名	烷烃	英文名
十三烷	tridecane	三十二烷	dotriacontane	六十烷	hexacontane
十四烷	tetradecane	三十三烷	tritriacontane	七十烷	heptacontane
十五烷	pentadecane	三十四烷	tetratriacontane	八十烷	Octacontane
十六烷	hexadecane	三十五烷	pentatriacontane	九十烷	nonacontane
十七烷	heptadecane	三十六烷	hexatriacontane	一百烷	hectane
十八烷	octadecane	三十七烷	heptatriacontane		
十九烷	nonadecane	三十八烷	octatriacontane		

3.2.2　烷基的命名

烷基的命名原则用后缀“**-yl**”取代相应烷烃后缀“**-ane**”，如下：

烷烃(alk**ane**) ⟷ 烷基(alk**yl**)

甲烷(meth**ane**) ⟷ 甲基(meth**yl**)

烷烃中通常将直链烷烃称为正烷烃(normal alkane)，正烷烃用“n-”表示，而“n-”通常可省略不写。例如：

n-dec**ane**(或 dec**ane**)　正癸烷(或癸烷)

n-but**ane**(或 but**ane**)　正丁烷(或丁烷)

n-hex**yl**(或 hex**yl**)　正己基(或己基)

n-undec**yl**(或 undec**yl**)　正十一烷基(或十一烷基)

采用普通命名的支链烷烃中还有四个常用的结构前缀“iso-”“sec-”“tert-”“neo-”，分别表示“异”“仲”“叔”“新”。例如：

$CH_3CH(CH_3)CH_2-$

iso-butyl
异丁基

$CH_3CH(CH_3)CH_2CH_2CH_3$

iso-hexane
异己烷

$CH_3CH_2CH(CH_3)-$

sec-butyl
仲丁基

$(CH_3)_3C-$

tert-butyl
叔丁基

$CH_3-C(CH_3)_2-CH_2-$

neo-pentyl
新戊基

$CH_3-C(CH_3)_2-CH_3$

neo-pentane
新戊烷

3.2.3 含有支链的烷烃的命名

含有支链烷烃的命名原则为“位号-烷基+烷烃”(*n*-radical+alkane)。例如:

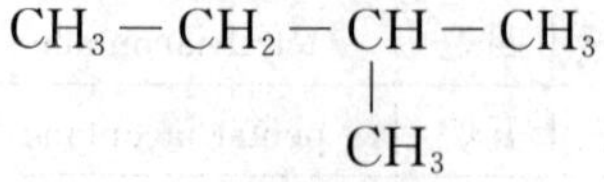

$$
\begin{array}{c}
CH_3-CH_2-\underset{\displaystyle \overset{|}{CH_3}}{CH}-CH_3
\end{array}
$$

2-methylbutane
2-甲基丁烷

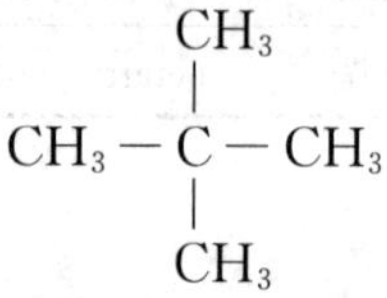

$$
CH_3-\underset{\displaystyle \overset{|}{CH_3}}{\overset{\displaystyle CH_3}{\overset{|}{C}}}-CH_3
$$

2,2-dimethylpropane
2,2-二甲基丙烷

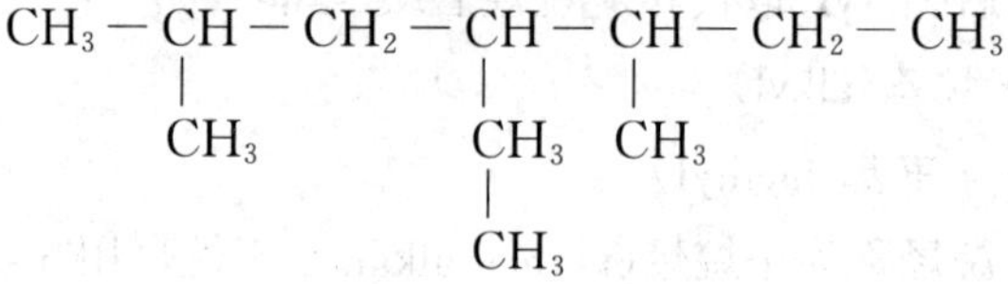

$$
CH_3-\underset{\displaystyle \overset{|}{CH_3}}{CH}-CH_2-\underset{\displaystyle \overset{|}{CH_2}-CH_3}{CH}-\underset{\displaystyle \overset{|}{CH_3}}{CH}-CH_2-CH_3
$$

4-ethyl-2,5-dimethylheptane
4-乙基-2,5-二甲基庚烷

结构较复杂的烷烃不能用普通命名法命名,只能采用系统命名法。选最长的碳链为主链,按相应的直链烷烃命名规则,从一端向另一端编号,支链作为取代基放在母体名称前,编号时使支链的编号尽可能小且支链按基团的字母顺序排列。

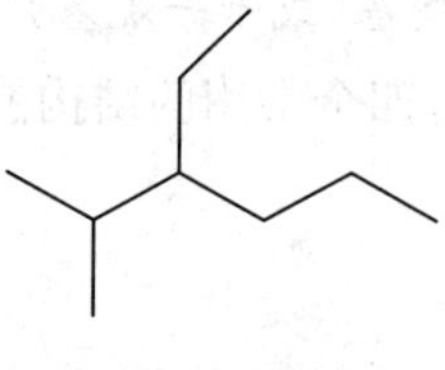

3-ethyl-2-methylhexane
3-乙基-2甲基己烷

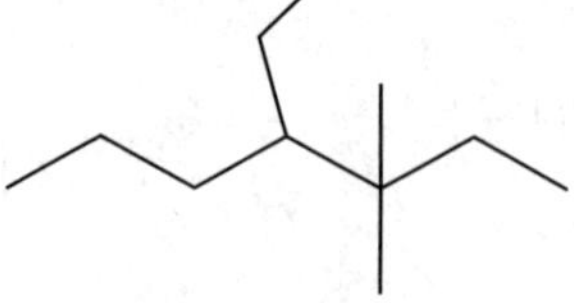

4-ethyl-3,3-dimethylheptane
4-乙基-3,3-二甲基庚烷

复杂的烷烃命名时须注意分子中有两个等长碳链时,按以下原则进行命名:

(1) 带支链数目较多者为主链。

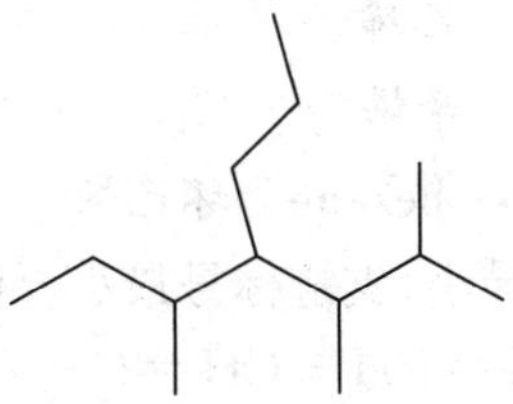

2,3,5-trimethyl-4-propylheptane
2,2,5-三甲基-4-丙基庚烷

(2) 支链定位号较小者为主链。

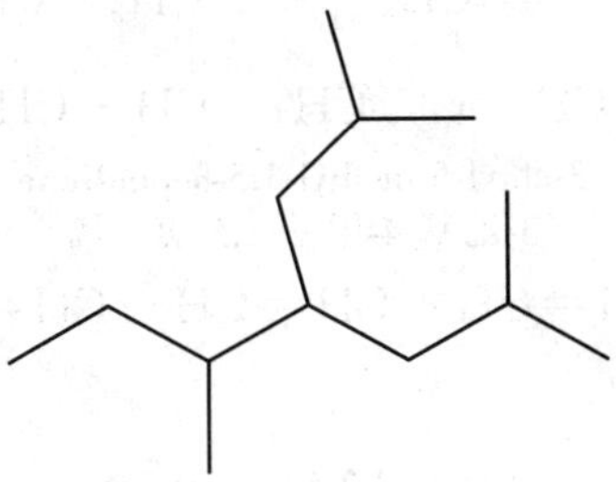

4-isobutyl-2,5-dimethylheptane
4-异丁基-2,5-二甲基庚烷

【练习3.1】 按系统命名法用英语命名下列烷烃。

$$\begin{array}{l} CH_3-CH_2-\underset{\displaystyle CH_3}{\underset{|}{CH}}-CH_3 \end{array} \qquad CH_3-\underset{\displaystyle CH_3}{\underset{|}{CH}}-CH_2-\underset{\displaystyle CH_2-CH_3}{\underset{|}{CH}}-CH_2-CH_3$$

$$\begin{array}{c} CH_3 \\ | \\ CH_3-CH_3 \\ | \\ CH_3-CH_2-CH_2-CH-CH-CH_2-CH_2-CH_2-CH_3 \\ | \\ CH_2-CH_3 \end{array}$$

$$\begin{array}{c} CH_3 \\ | \\ CH_3-C-CH_3 \\ | \\ CH_3-CH_2-CH-CH_2-CH_2-CH_3 \end{array}$$

3.3　烯烃和炔烃的命名

3.3.1　烯烃的命名

烯烃(alkene)的命名原则是将相应烷烃的后缀“**-ane**”改为“**-ene**”，其形式为“数字前缀-**ene**”(number prefix-**ene**)。例如：

ethane　乙烷　　ethene　乙烯

octane　辛烷　　octene　辛烯

cyclohexane　环己烷　　cyclohexene　环己烯

含有支链的多烯烃:双键标号最小,支链标号最小。例如:

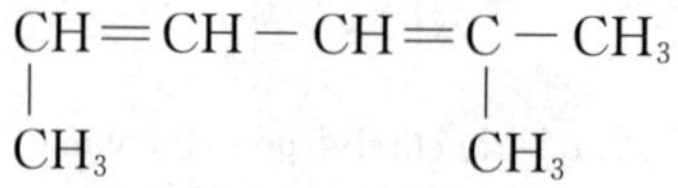

$$\begin{array}{l}CH=CH-CH=\underset{|}{C}-CH_3\\ \! | \\ CH_3 \qquad\qquad\quad CH_3\end{array}$$

2-methyl-2,4-hexa**diene**
2-甲基-2,4-己二烯

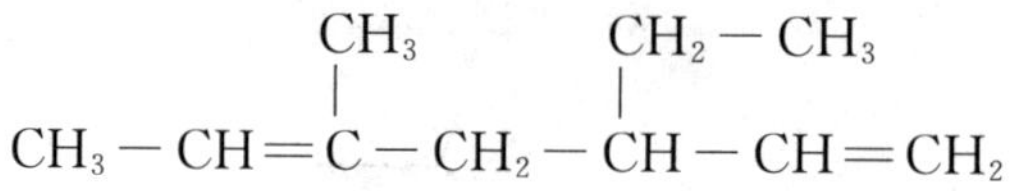

$$CH_3-CH=\overset{\overset{CH_3}{|}}{C}-CH_2-\overset{\overset{CH_2-CH_3}{|}}{CH}-CH=CH_2$$

3-ethyl-5-methyl-1,5-hept**adiene**
3-乙基-4-甲基-1,5-庚二烯

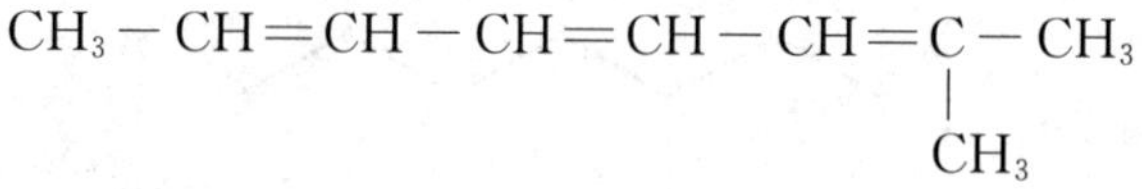

$$CH_3-CH=CH-CH=CH-CH=\underset{\underset{CH_3}{|}}{C}-CH_3$$

2-methyl-2,4,6-oc**tatriene**
2-甲基-2,4,6-辛三烯

3.3.2　炔烃的命名

炔烃(alkyne)的英文命名原则为“数字前缀**-yne**”(number prefix**-yne**),即将相应烷烃的后缀“**-ane**”改为“**-yne**”。例如:

propane　丙烷　　propyne　丙炔

butane　丁烷　　butyne　丁炔

3.3.3　烯炔的命名

烯烃(alkenyne)的英文后缀如下:

一烯一炔　-en-*n*-yne

二烯一炔　-adien-*n*-yne

三烯一炔　-atrien-*n*-yne

一烯二炔　-en-*n*-diyne

其中,*n* 为炔的位次。例如:

$$C\equiv C-CH=CH-CH_3$$

3-**penten-1-yne**
3-戊烯-1-炔

$$C\equiv C-CH=CH-CH=CH_2$$

1,3-hex**adien-5-yne**
1,3-己二烯-5-炔

$$
\begin{array}{l}
C \equiv C - C \equiv C \\
\qquad\qquad\quad | \\
\qquad\qquad CH_2 - CH_2 - CH = CH_2
\end{array}
$$

7-octene-1,3-diyne
7-辛烯-1,3-二炔

系统命名中,应选含不饱和键最多且最长的直链为主链。如分子中有两条直链具有相同数目的不饱和键时,取碳原子数较多者为主链。如碳原子数相同,则取含双键数目较多者为主链。例如:

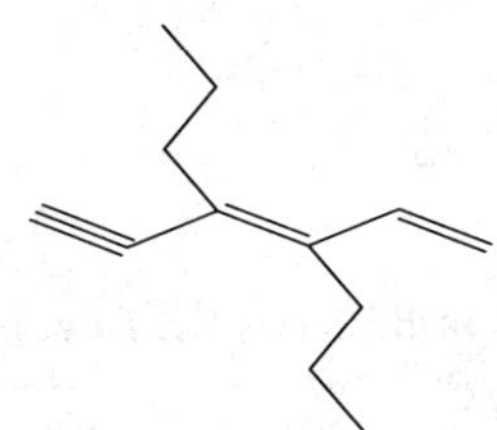

3,4-dipropyl-1,3-hexadien-5-yne
3,4-二丙基-1,3-己二烯-5-炔

3.3.4　不饱和烃基的命名

不饱和烃基的命名类似于饱和烃基,但需要标出不饱和键的位次。其中,烯基的构成,用词尾“**-yl**”取代相应烯烃最后字母“**e**”;炔基的构成,用词尾“**-yl**”取代相应炔烃最后字母“**e**”。如下:

烯烃 alkene ⟷ 烯基 alken**yl**

炔烃 alkyne ⟷ 炔基 alkyn**yl**

$CH_3-CH=CH-$	1-丙烯基(1-propenyl)/丙烯基
$CH_2=CH-CH_3=CH-$	1,3-丁二烯基 1,3-butadienyl
$C\equiv C-$	乙炔基(ethynyl)
$C\equiv C-CH_2-$	2-丙炔基(2-propynyl)
$C\equiv C-CH=CH-CH_2-$	2-戊烯-4-炔基

有些简单的不饱和烃基可用俗名。例如:

$CH_2=CH-$	乙烯基(vinyl)
$CH_2=CH-CH_2-$	烯丙基(allyl)
$CH_2=C(CH_3)-$	异丙烯基(isopropenyl)

3.3.5　多价基的命名

双键在同一个碳原子上称为“亚”,英文词尾为“亚基”(**-ylidene**)。例如:

$CH_2=$	亚甲基(methylene [ˈmeθɪliːn])
$CH_3CH=$	亚乙基(eth**ylidene** [əˈθilidiːn])
$(CH_3)_2C=$	亚异丙基(isopropy**lidene** [ˌaisəuprəuˈpiləˌdiːn])

三键在同一个碳原子上的称为次基,英文词尾为"**-yl-idyne**"。例如:

$CH_3-C\equiv$ 次乙基(ethyl**idyne**)

$CH\equiv$ 次甲基(methyl**idyne**)

3.4 环烃的命名

3.4.1 饱和单环烃的命名

饱和单环烃(monocyclic hydrocarbon)的命名原则为"**cyclo**-相应烷烃名"(**cyclo**-alkane)。例如:

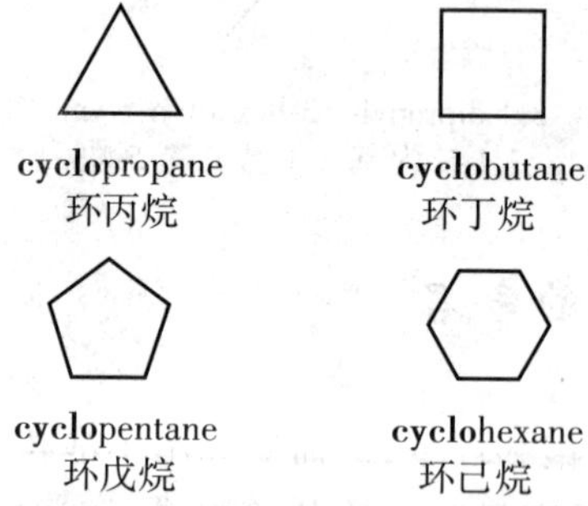

3.4.2 含取代基的脂肪环烃的命名

环上带有侧链时,如果侧链的碳原子数比环内的碳原子数少,则侧链作为取代基。如果侧链的碳原子数目等于或多于环内碳原子数,或侧链有不止一个脂环时,则将环作为取代基命名。例如:

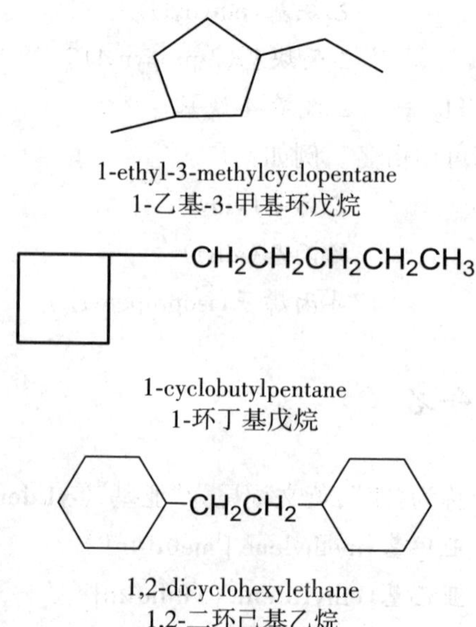

3.4.3　环烷基的命名

环烷基的命名与烷基相同，用词尾“**-yl**”取代相应烷烃词尾“**-ane**”。例如：

环烷烃(cycloal**kane**)⟷ 环烷基(cycloal**kyl**)

2,4-cyclopentadien-1-**yl**
2,4-环戊二烯-1-基

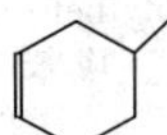

3-cyclohexene-1-**yl**
3-环己烯-1-基

CH_3

2-methyl-2,4-cyclohexadiene-1-**yl**
2-甲基-2,4-环己二烯-1-基

3.4.4　不饱和单环烃的命名

不饱和单环烃的命名是把相应的饱和单环烃的词尾“-ane”改为“-ene”(烯)、“-yne”(炔)、“-adiene”(二烯)、“-adiyne”(二炔)、“-enyne”(烯炔)等，并使不饱和键尽可能取最小编号。

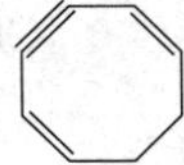

1,5-cyclooctadien-3-yne
1,5-环壬二烯-3-炔

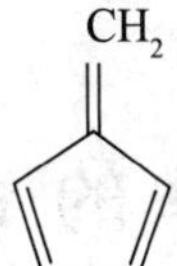

5-methylene-1,3-cyclopentadiene
5-亚甲基-1,3环戊二烯

3.5　芳烃的命名

在单环芳烃(aromatic hydrocarbon)中，当苯环上连有烃基时，苯环和烃基都可作为母体，主要由烃基的大小决定。当两个或更多的苯环连在同一个碳原子上或碳链上时,可将苯环作为取代基命名。

两个烷基取代的苯环,因为取代位置不同,可以有三个异构体,可用阿拉伯数字表示,也可用"邻""间""对"表示,英文名称则分别用"o-"(ortho-)、"m-"(meta-)、"p-"(para-)表示。

当苯环上联有三个取代基时,由于它们的位置不同而常用数字位号区别,若取代基是相同的,则可用"连"(vic-)、"偏"(unsym-)、"均"(sym-)来表示。例如:

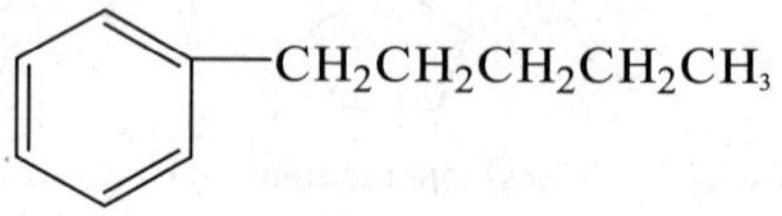

$C_6H_5-CH_2CH_2CH_2CH_2CH_3$

pentylbenzene
正戊苯

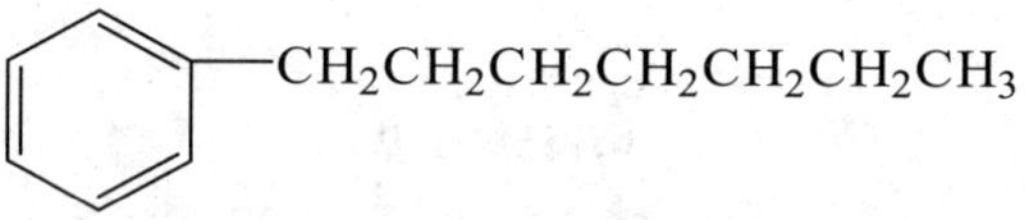

$C_6H_5-CH_2CH_2CH_2CH_2CH_2CH_2CH_3$

1-phenylheptane
庚基苯

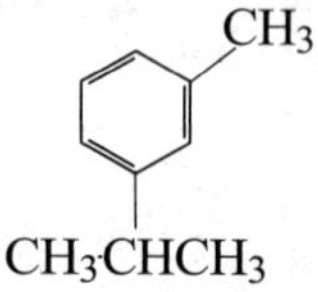

CH_3, CH_3CHCH_3

3-isopropyl-1-methylbenzene
3-异丙基-1-甲基苯

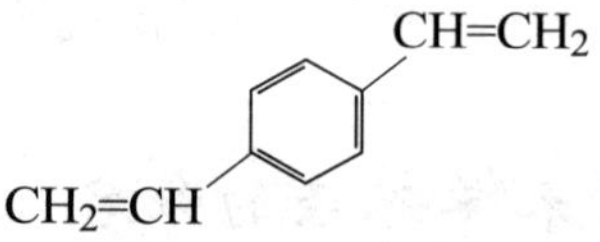

$CH{=}CH_2$, $CH_2{=}CH$

1,4-divinylbenzene
1,4-二乙烯基苯

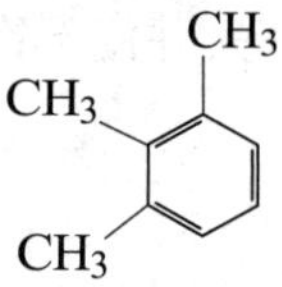

CH_3, CH_3, CH_3

1,2,3-trimethylbenzene
1,2,3-三甲基苯

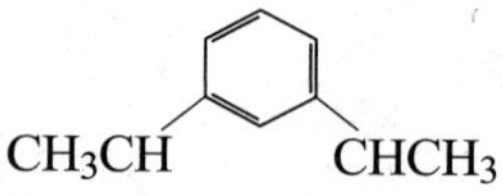

CH_3CH, $CHCH_3$

m-diethylbenzene
间-二乙基苯

3.6 醇的命名

3.6.1 一元醇的命名

一元醇在命名时,通常将相应烃的名称的最后一个字母"e"用"-ol"代替。例如:

methane 甲烷	methan**ol** 甲醇
cyclohexane 环己烷	cyclohexan**ol** 环己醇
2-propene 2-丙烯	2-propen**ol** 2-丙烯醇
3-methyl-1-pentyne 3-甲基-1-戊炔	3-methyl-1-pentyn-3-alch**ol** 3-甲基-1-戊炔-3-醇

3.6.2　二元醇的命名

二元醇的名称构成是在相应烃的后面加上二元醇词尾“-diol”。例如：

butane　丁烷　　1,4-butane**diol**　1,4-丁二醇

2,4-hexadiyne　2,4-己二炔　　2,4-hexadiyne-1,6-**diol**　2,4-己二炔-1,6-二醇

cyclohexane　环己烷　　1,4-cyclohexane**diol**　1,4-环己二醇

3.6.3　烷氧基和醇盐的命名

烷氧基(alkoxy)的命名原则是用词尾“-oxy”代替相应烃基词尾“-yl”：

alk**yl**(烃基)⟷alk**oxy**(烷氧基)

例如：

eth**yl**　乙基　　eth**oxy**　乙氧基

cyclohex**yl**　环己基　　cyclohex**oxy**　环己氧基

醇盐(alkoxide)的名称由金属名称 + 烷氧化合物(alkoxide)名称构成，而烷氧化合物名称是用词尾“-oxide”代替相应烃基词尾“-yl”：

烷基(alk**yl**) ⟷ 烷氧化合物(alk**oxide**)

例如：

eth**yl**　乙基　　eth**oxide**　乙氧化物

sodium eth**oxide**　乙醇钠　　aluminum isoprop**oxide**　异丙醇铝(或三异丙氧基铝)

3.7　醚和酚的命名

3.7.1　醚的命名

醚(ether)的命名原则是分别写出两个烃基名称再加上“ether”或用烷氧基(alkoxy)作为取代基命名。例如：

ethyl ether　乙醚　　ethyl methyl ether　甲(基)乙(基)醚

再如 $CH_3CH_2-O-CH=CH_2$ 可以命名为 ethyl ethenyl ether(乙基乙烯基醚)或 ethoxyethene(乙氧基乙烯)。

3.7.2　酚的命名

酚以苯酚(phenol)作为母体来命名，例如：

2-methyl-4-ethenylphenol　2-甲基-4-乙烯基苯酚

3.8 醛和酮的命名

3.8.1 醛的命名

醛(aldehyde)的命名原则是将相应烃的最后一个字母"e"用词尾"-al"代替,例如:

methane　甲烷　　methan**al**　甲醛

2-butene　2-丁烯　　2-buten**al**　2-丁烯醛

2-methyl-3-pentyne　2-甲基-3-戊炔　　2-methyl-3-pentyn**al**　2-甲基-3-戊炔醛

3.8.2 酮的命名

酮(ketone)的命名原则是将相应烃的最后一个字母"e"用词尾"-one"代替。例如:

butane　丁烷　　butan**one**　丁酮

5-ethyl-3-heptene　5-乙基-3-庚烯　　5-ethyl-3-heptene-2-**one**　5-乙基-3-庚烯-2-酮

3.8.3 二醛、二酮的命名

二醛的命名原则是在相应烃的后面加上词尾"-dial";二酮的构成是在相应烃的后面加上词尾"-dione"。例如:

ethane　乙烷　　ethane**dial**　乙二醛

butane　丁烷　　2,3-butane**dione**　2,3-丁二酮

3.9 羧酸及其衍生物的命名

3.9.1 羧酸的命名

一元羧酸用词尾"-oic"代替相应烃中最后一个字母"e"然后加上单词"acid";二元羧酸将词尾"-dioic"加在相应烃的最后,然后加上单词"acid"。例如:

methane　甲烷　　methan**oic** acid　甲酸

propyne　丙炔　　propyn**oic** acid　丙炔酸

butane　丁烷　　butanedi**oic** acid　丁二酸

2,6-octadien-4-yne　2,6-辛二烯-4-炔　2,6-octadien-4-yne**dioic** acid　2,6-辛二烯-4-炔二酸

3.9.2 羧酸酯和羧酸盐的命名

羧酸根的命名原则是用词尾"-ate"代替相应羧酸中的词尾"-oic",再去掉单词"acid",例如:

methan**oic** acid　甲酸　　methano**ate**　甲酸根

羧酸酯(ester或carboxylate)和羧酸盐(carboxylate)的英文名称分别为"烃基 + 羧酸根"和"金属或铵根 + 羧酸根"。例如:

ethyl ethanoate　乙酸乙酯　　dimethyl-1,4- butanedioate　1,4-丁二酸二甲酯

sodium ethanoate　乙酸钠　　ammonium methanoate　甲酸铵

3.9.3 酰卤的命名

酰基(acyl)的命名原则是用词尾"-yl"代替相应羧酸中的词尾"-oic",再去掉单词"acid"。例如:

ethan**oic** acid　乙酸　　ethano**yl**　乙酰基

酰氯和酰溴的命名原则分别是"acyl(酰基) + chloride"和"acyl(酰基) + bromide"。例如:

ethanoyl chloride　乙酰氯　　butanoyl bromide　丁酰溴

3.9.4 酰胺的命名

酰胺(amide)的命名原则是用词尾"-amide"代替相应羧酸中的词尾"-oic",再去掉单词"acid"。例如:

methan**oic** acid　甲酸　　methan**amide**　甲酰胺

propen**oic** acid　丙烯酸　　propen**amide**　丙烯酰胺

3.9.5 酸酐的命名

酸酐(anhydride)的命名原则是用"anhydride"代替相应羧酸的"acid"。例如:

ethanoic　acid　乙酸　　ethanoic　anhydride　乙酸酐

methanoic　ethanoic　anhydride　甲(酸)乙(酸)酐

【练习3.2】 写出下列常见化学基团和英文前缀的中文名称。

acyl　　benzyl

amino　　bromo

alkyl　　carbonyl

carboxy(l)	i(iso)-
chloro	m(meta)-
cyano	nitro
fluoro	nitroso
formyl	o(ortho)-
halo	p(para)-
hydroxy	phenyl
imino	s(sec)-
iodo	t(tert)-

材料阅读

Organic Nomenclature

Organic nomenclature is the system established for naming and grouping organic compounds.

Formally, rules established by the international union of pure and applied chemistry (known as IUPAC nomenclature) are authoritative for the names of organic compounds, but in practice, a number of simply-applied rules can allow one to use and understand the names of many organic compounds.

For many compounds, naming can begin by determining the name of the parent hydrocarbon and by identifying any functional groups in the molecule that distinguish it from the parent hydrocarbon. The numbering of the parent alkane is used, as modified, if necessary, by application of the Cahn Ingold Prelog priority rules in the case that ambiguity remains after consideration of the structure of the parent hydrocarbon alone. The name of the parent hydrocarbon is modified by the application of the highest-priority functional group suffix, with the remaining functional groups indicated by numbered prefixes, appearing in the name in alphabetical order from first to last.

In many cases, lack of rigor in applying all such nomenclature rules still yields a name that is intelligible—the aim, of course, being to avoid any ambiguity in terms of what substance is being discussed.

For instance, strict application of CIP priority to the naming of the compound

$$NH_2CH_2CH_2OH$$

would render the name as 2-aminoethanol, which is preferred. However, the name 2-hydroxyethanamine unambiguously refers to the same compound.

How the name was constructed:

(1) There are two carbons in the main chain; this gives the root name “eth”.

(2) Since the carbons are singly-bonded, the suffix begins with "an".

(3) The two functional groups are an alcohol (OH) and an amine (NH_2). The alcohol has the higher atomic number, and takes priority over the amine. The suffix for an alcohol ends in "ol", so that the suffix is "anol".

(4) The amine group is not on the carbon with the OH (the #1 carbon), but one carbon over (the #2 carbon); therefore we indicate its presence with the prefix "2-amino".

(5) Putting together the prefix, the root and the suffix, we get "2-aminoethanol".

There is also an older naming system for organic compounds known as common nomenclature, which is often used for simple, well-known compounds, and also for complex compounds whose IUPAC names are too complex for everyday use.

Simplified molecular input line entry specification (SMILES) strings are commonly used to describe organic compounds, and as such are a form of naming them.

(摘自　http://www.chemistrydaily.com/chemistry/Organic_nomenclature.)

有机化学常用词汇

acid anhydride　酸酐
acyl halide　酰卤
alcohol　醇
aldehyde　醛
aliphatic　脂肪族的
alkane　烷烃
alkene　烯烃
alkyne　炔
allyl　烯丙基
amide　氨基化合物
amine　胺
amino acid　氨基酸
aromatic compound　芳香化合物
aromatic ring　芳环,苯环
branched-chain　支链
butyl　丁基
carbonyl　羰基
carboxyl　羧基
chain　链
chelate　螯合
chiral center　手性中心
conformers　构象
copolymer　共聚物
derivative　衍生物
dextrorotatory　右旋性的
diazotization　重氮化作用
dichloromethane　二氯甲烷
ester　酯
ethyl　乙基
fatty acid　脂肪酸
functional group　官能团
General formula　通式
Glycerol　甘油,丙三醇
heptyl　庚基
heterocyclic　杂环的
hexyl　己基
homolog　同系物
Hydrocarbon　烃,碳氢化合物
hydrophilic　亲水的
hydrophobic　疏水的

hydroxide 羟基
ketone 酮
levorotatory 左旋性的
methyl 甲基
molecular formula 分子式
monomer 单体
octyl 辛基
Open chain 开链
optical activity 旋光性(度)
organic 有机的
organic chemistry 有机化学
Organic compounds 有机化合物
pentyl 戊基
phenol 苯酚
phenyl 苯基
Polymer 聚合物,聚合体
propyl 丙基
ring-shaped 环状结构
saturated compound 饱和化合物
side chain 侧链
straight chain 直链
Structural formula 结构式
Tautomer 互变(异构)体
triglyceride 甘油三酸酯
unsaturated compound 不饱和化合物

第2篇

翻译与阅读

第4章 化学化工专业英语翻译

4.1 翻译标准

化学化工专业英语的翻译标准有四条:准确、通顺、简练和符合专业用语习惯。这要求译者必须准确理解和掌握原文的内容,不能仅凭个人的想法推测。在忠实于原文内容和风格的基础上,译者还必须运用汉语把原文通顺、简练地表达出来,同时注意表达方式要符合专业用语习惯。例如:

Heat-treatment is used to normalize, to soften or to harden steels.

译文1:热处理被用来使钢正常化,软化或硬化。(误译)

译文2:热处理可用来对钢进行正火、退火或淬火。(正译)

normalize、soften和harden在普通英语中有“正常化”“软化”和“硬化”的意思。

The homologs of benzene are those containing an alkyl group or alkyl groups in place of one or more hydrogen atoms.

译文1:苯的同系物就是那些被一个或多个烷基取代一个或多个氢原子所形成的产物。(误译)

译文2:苯的同系物是苯环上那些含有单烷基(取代一个氢)或多烷基(取代多个氢)的物质。(正译)

译文1所表达的意思是基本正确的,但是语句啰嗦,读起来令人费解;而译文2则通顺简练,易于理解。

4.2 翻译过程

翻译前,先要通读全文,划出生词及疑难点。通过查阅工具书解决理解上的困难后,完整、准确地理解全文,然后再逐句加以分析、翻译。译出的文字要按照汉语的表达习惯加以修改、组织。最后对文字进行推敲、加工,力求译文真实、通顺。在进行句子的翻译时,可以借助一些方法或技巧。

4.2.1　理解阶段

理解原文是翻译的基础和关键。译者不仅要弄清原文中每个单词、词组和句子的准确含义,还要弄清整个篇章的结构、句子的逻辑、与上下文的关系、所采用的语气等,即掌握原文的全部精神实质。不仅要明白字面上的意思,还要理解其内在的信息和意图。看下面这段原文:

It takes time before ozone-depleting chemicals can be totally eliminated. First, industries must develop harmless ways to replace the CFCs we use for important purpose like refrigeration. Also, some countries do not have the means to use substitutes. So a special fund has been set up to help these nations use new chemicals and technology.

在翻译这段英文时,要注意单词"means"的含义。"means"有两个常见的意思,一是"方式、方法",二是"收入、财力"。那么在上面这段话中,"means"到底应该翻译成什么？我们知道,一段文字是一个有机整体,词与词、词与句子、句子与段落甚至整个篇章之间,都有着必然的内在联系。单从句子来看,这里的"means"译成"方式,方法"似乎可行,译为"同样,一些国家也没有使用替代品的方法"。但是结合上下文来看,上文中"完全消除消耗臭氧的化学品还需要一段时间。因为,首先,产业界必须开发无害的途径来取代被用于重要目的如制冷剂的氟利昂",在这个句子中已经提到了"开发途径(方法)",那么下一个句子中,很显然不会重复同一个理由。而且,下文的"So a special fund has been set up to help these nations use new chemicals and technology."(一个特殊基金会被建立起来帮助这些国家使用新的化学品和技术)与本句为因果关系,根据上下文,很容易推断出,本句中的"means"应该译为"收入、财力",正因为"一些国家没有使用替代品的财力",因此,"需要建立特殊基金会来帮助它们"。

由此可见,要想准确理解化工英语文章,不仅需要弄懂每个单词的意思,还要把单词放入整个语篇和语境下考量,才能掌握原文的全部精神实质。

4.2.2　表达阶段

表达就是用汉语把已经理解了的原文内容用明白、晓畅的方式叙述出来。在表达阶段,要注意语言的规范性和逻辑性。表达得好坏取决于对原文理解的程度和对汉语的运用程度。

4.2.3　校对阶段

校对阶段是对理解和表达的内容进行校对,检查译文是否能准确无误地转述原作内容,译文的语言表达是否规范、是否符合汉语习惯,并通过进一步修改、润色,形成最佳的译文。

4.3 直译与意译

直译和意译是翻译过程中两种最常见的翻译方法。直译是指既保持原文内容，又保持原文形式的翻译方法。直译采用原文的表现手段，对句子结构和语序不作调整或不作重大调整。意译是指只保持原文内容，不保持原文形式的翻译方法。意译的译文采用新的等效的表现手段，句子结构有可能做较大的调整。直译与意译相互关联、互为补充，同时，它们又互相协调、互相渗透，不可分割。

4.3.1 直译

直译指不仅忠实于原文内容，而且忠实于原文形式的翻译方法。这个定义包含以下三个方面的含义：① 忠于原文内容；② 忠于原文形式，要求在保持原文内容的前提下，力求使译文与原文在选词用字、句法结构、形象比喻及风格特征等方面尽可能相同；③ 译文形式通顺流畅。例如：

pillar industry 支柱产业　　acid rain 酸雨

While the molecules in crude oil include different atoms such as sulfur and nitrogen, the hydrocarbons are the most common form of molecules.

尽管原油分子中含有硫和碳这样的原子，但烃是最常见的分子形式。

However, forethought and careful wording can minimize these hazards.

然而，事先考虑和小心工作可以尽可能降低危害。

When it is combined with sand (SiO_2) and calcium carbonate ($CaCO_3$) and heated to very high temperatures, then cooled very rapidly, glass is produced.

将碳酸钠与沙子、碳酸钙混合在一起，加热至高温，并快速冷却，就产生了玻璃。

初学翻译的人可能会逐字逐句地翻译，认为这就是“忠实”于原文。但由于英语和汉语在句法结构和行文用词上有很大的区别，如果不考虑中英文的区别，单纯地以词为单位，机械地将译文的词汇与原文对等，过分拘泥于原文形式，就会导致译文生硬晦涩、难以理解。这种翻译方法我们称为“死译”。例如：

Manganese has the same effect on the strength of steel as silicon.

锰有同样影响在强度上像硅。

这个句子既不忠实于原文，又不符合汉语表达习惯，属于硬照着原文逐字翻译地“死译”，应该通过调整译为“锰像硅一样会影响钢的强度”。

4.3.2　意译

意译也称为“自由翻译”，指忠实于原文内容，但是不拘泥于原文的结构形式和修辞手法的翻译方法。这个定义包含以下三个方面的含义：① 忠实于原文内容；② 在保持原文内容的前提下，力求使译文与原文在选词用字、句法结构、形象比喻及风格特征等方面符合读者的阅读习惯；③ 译文形式通顺流畅。例如：

It is to be hoped that the next century will witness a wider national distribution of the Laureates.

人们希望在下个世纪，更多国家的学者们能获得诺贝尔奖。

We can get more current from cells connected in parallel.

电池并联时提供的电流更大。

需要注意的一点是，意译虽然可以改变原文的形式和结构，但不能歪曲原文的意思，随意增减原文的内容进行所谓的“自由翻译”，这样就成了“胡译”。例如：

The first author (LAC) would like to express his gratitude to the late Dr. Alf de Ruvo for introducing me to sandwich structures as a graduate student and the many years of fruitful cooperation on sandwich structures at SCA research.

原译：作者也非常高兴地向阿尔夫德鲁沃博士表达晚到的谢意，感谢在我研究生期间领我进入夹层结构领域以及随后多年来在夹层结构SCA研究领域的富有成效的合作。

这个译文中，作者自己加上了原文中没有出现的“高兴”一词，而且理解错了“late”一词的意思。同样地，作者也曲解了“SCA research”中的“SCA”的意思，这并不是某一个领域的名称缩写。

直译与意译之间没有一个确切的界限，不管使用哪种方法，都要注意原文的创作意图和思想内容是否完整地表达出来，译文的语言是否规范，是否能为译语读者所接受。如果不能遵循这一原则，过分注重形式对等或曲解原文含义，都会造成“死译”或“胡译”。

4.4　词的增译和省译

词的增译是指基于上下文的意思、逻辑关系以及汉语的表达习惯，增加词量，以表达原文字面没有出现但实际已经包含的意思。词的省译是指在英语原句中，有的单词从语法结构上来说是必不可少的，但又没有什么实际意义，只是在原句中单纯地起着语法作用；有的单词虽然有实际意义，但按照字面意思译出又显得多余、累赘，这样的单词在翻译时往往可以省略不译。

简而言之，词的增译和省译是根据汉语表达习惯，在译文中增加一些原文中没有的词或减去原文中多余的词。

4.4.1 词的增译

词的增译遵循以下两个方面的原则：

1. 增补语法上需要的词

英语中的可数名词有单数和复数之分，但汉语中的名词没有单、复数的区别，因此在译文中要根据汉语表达习惯，增加合适的量词。例如：

The first electronic computers went into operation in 1945.

第一批电子计算机于1945年投入使用。

英语中常用省略句，译文要根据汉语的表达习惯，把省略的部分译出来。例如：

Matter can be changed into energy, and energy into matter.

物质可以转化为能量，能量也可以转化为物质。

英语通过改变动词词尾或使用助动词可以构成不同的时态，但汉语动词没有时态变化，而用时间副词来表示时态，如用“现在”“将来”“曾经”表示不同的时间，用“正在”“已经”“经常”等时间或频率副词及“着”“了”等助词表示时间的进程或状态。例如：

Organic compounds were once thought to be produced only by living organism.

以前曾认为有机化合物只能在有生命的机体中产生。

2. 增补语义上需要的词

英语中的某些抽象名词和不及物动词，若单独译出，表达不够清晰完整，通常在这些词的后面加上“状态”“作用”“过程”“情况”等词。例如：

Oxidation will make iron and steel rusty.

氧化作用会使钢铁生锈。

为了使译文前后连贯、意思通顺、逻辑严密，可以适当增补一些词，如概括词、连词等。例如：

According to scientists, it takes nature 500 years to create an inch of topsoil.

根据科学家的看法，自然界要用500年的时间才能形成一英寸厚的表层土壤。

Heat from the sun stirs up the atmosphere, generating winds.

太阳发出的热能搅动大气，于是产生了风。

当句中有几个成分并列时，可以根据并列成分的数量增加数词和量词，如“个”“者”“方面”“因素”等。例如：

The frequency, wave length and speed of sound are closely related.

频率、波长和声速三者是密切相关的。

The resistance of the pipe to the flow of water through it depends upon the length of the pipe, and the feature of the inside walls (rough or smooth).

水管对通过的水流的阻力取决于三个因素，分别为管道长度、管道直径、管道内壁的特性(粗糙或光滑)。

【练习4.1】 将下列各句翻译成汉语。

(1) Using a transformer, power at low voltage can be transformed into power at high voltage.

译:

(2) We made the transistors by different means only to get the same effect.

译:

(3) The frequency, wave length and speed of sound are closely related.

译:

(4) Atomic cells are very small and very light, as compared to ordinary dry ones.

译:

(5) The cost of such a power plant is a relatively small portion of the total cost of the development.

译:

(6) Being stable in air at ordinary, mercury combines with oxygen if heated.

译:

(7) Because of its reliability and simplicity evaporation is often applied on a large scale, normally using steam as the heat source.

译:

(8) The frequency, wave length, and speed of sound are closely related.

译:

(9) The virus may survive weeks and months.

译:

(10) The lion is the king of animals.

译:

(11) The revolution of the earth around the sun causes the changes of the seasons.

译:

(12) The missiles are skimming over the sea.

译:

(13) Circling venus, the spaceship took a lot of photographs.

译:

4.4.2 词的省译

有些词语或句子成分在英语中是必不可少的,但如果一律加以照搬,就会影响汉语译文的简洁和通顺。因此,为了使译文简练且更加符合汉语的表达习惯,需要省略一些可有可无或译出来反显累赘的词语。词的省译包含以下6种类型:

1. 省略冠词

英语有冠词而汉语没有,因此,英译汉时往往将冠词省略。例如:

Ice is a solid. If we heat it, it melts and becomes water.
冰是固体,如果加热,就会融化成水。

The atom is the smallest particle of an element.
原子是元素的最小粒子。

2. 省略代词

英语中代词使用的较多,而汉语中却较少使用代词。例如:

If you know the frequency, you can find the wave length.
知道频率,就能求出波长。

A gas distribute itself uniformly throughout a container.
气体均匀地分布在整个容器中。

3. 省略动词

英语里的联系动词以及一些抽象的行为动词翻译时可以省略。例如:

Stainless steels possess good hardness and high strength.
不锈钢硬度大、强度高。

4. 省略名词

The world of work injury insurance is complex.
工伤保险十分复杂。

5. 省略连词

A glass becomes hotter if it is compressed.
玻璃被压缩后会变热。

6. 省略介词。

The molecular structure is different for various kinds of polymers.

各种聚合物的分子结构不同。

【练习4.2】 将下列各句翻译成汉语。

(1) In the absence of force, a body will either remain at rest, or continue to move with constant speed in a straight line.

译:

(2) All substance is made of atoms whether it is a solid, a liquid, or a gas.

译:

(3) The products should be sampled to check their quality before they leave the factory.

译:

(4) In figure 5, *A* is the resistance, *L* is the distance and *C* is the effort.

译:

(5) The interfacial tension between the two phases is significantly lower than that in water-organic solvent systems.

译:

(6) The Pacific alone covers an area larger than that of all the continents put together.

译:

4.5 词类转译法

词类转换法是指在翻译过程中,由于英、汉两种语言的结构与表达方式不同,常常需要把原文中属于某种词类的词,根据译文的语言习惯进行转换,译成另一种词类,从而使译文通顺自然的翻译方法。

4.5.1 名词的转译

1. 转译成形容词

In certain cases friction is an absolute necessity.

在某些情况下，摩擦是绝对必要的。

2. 转译成动词

The sight and sound of our jet planes filled me with special longing.

看到喷气式飞机，听见隆隆的声音，令我特别神往。

Can artificial fibers be used as substitutes for natural fibers?

人造纤维能代替天然纤维吗?

3. 转译成副词

Efforts to apply computer techniques have been a success in improving pyrolysis techniques.

计算机技术的应用已经成功地促进了热解技术的发展。

4.5.2 动词的转译

1. 转译成名词

The machine weighs about 500 kilograms.

这台机器的质量大约是500千克。

An acid and a base react in a proton transfer reaction.

酸碱反应是质子转移的反应。

2. 转译成形容词

This machine works efficiently.

这台机器的工作效率很高。

4.5.3 形容词的转译

1. 转译成名词

The cutting tool must be strong, tough, hard, and wear resistant.

刀具必须具有足够的强度、硬度、韧性和耐磨性。

Computers are more flexible, and can do a greater variety of jobs.

计算机的灵活性更大，因此能做更多种类的工作。

英语中，有些形容词加定冠词“the”后表示一类人(或物)，这类形容词可称为名词化的形容词。例如：

Electrons move from the negative to the positive.

电子由负极流向正极。

2. 转译成动词

They are quite content with the data obtained from the experiment.

他们对实验中获取的数据非常满意。

If we were ignorant of the structure of the atom, it would be impossible for us to study nuclear physics.

忽视原子的结构，将无法研究核物理学。

3. 转译成副词

The mechanical automatization makes for a tremendous rise in labor productivity.

机械自动化可以大大提高劳动生产率。

Below 4 ℃,water is in continuous expansion instead of continuous contraction.

水在4 ℃以下不断地膨胀，而非不断收缩。

4.5.4　副词的转译

1. 转译成名词

This crystal is dimensionally stable.

这种晶体的尺寸很稳定。

These parts must be proportionally correct.

这些零件的比例必须准确无误。

2. 转译成形容词

It is demonstrated that dust is extremely hazardous.

已经证实，粉尘具有极大的危害。

Sulphuric acid is one of extremely reactive agents.

硫酸是强烈反应试剂之一。

3. 转译成动词

The reaction force to this action force pushes the rocket ship along.

这个作用力所产生的反作用力推动宇宙飞船前进。

The chemical experiment is over.

化学实验结束了。

4.5.5　介词的转译

有动作意义的介词，如for、by、in、past、with、over、into、around、across、toward、through等，汉译时，可转换为汉语的动词。例如：

The type of film develops in twenty minutes.

冲洗此类胶卷需要20分钟。

The term laser stands for amplification by stimulated emission of radiation.
“激光”这个术语指的是利用辐射的受激发射放大光波。

【练习4.3】 将下列各句翻译成汉语。

(1) The law of thermodynamics is of great importance in the study of heat.

译:

(2) The output voltages of the control system varied in a wide range.

译:

(3) Generally speaking, methane series are rather inert.

译:

(4) This solar cell is only 7% efficient.

译:

(5) Both the compounds are acids, the former is strong, the latter weak.

译:

(6) As you know that also present in solid are numbers of free electrons.

译:

(7) He is quite familiar with the performance of this machine.

译:

(8) A helicopter is free to go almost anywhere.

译:

(9) Oxygen is one of the important element in the physical world, it is very active chemically.

译:

(10) If one generator is out of order, the other will produce electricity instead.

译:

(11) Rockets have found application for the exploration of the universe.

译:

(12) Neutrons act differently from protons.

译：

(13) With slight modifications each type can be used for all three systems.

译：

(14) We are all familiar with the fact that nothing in nature will either start or stop moving of itself.

译：

(15) The device is shown shematically in Figure 8.

译：

(16) This film is uniformly thin.

译：

4.6　成分转译法

句子成分转换是指词类不变而成分改变的译法。通过改变原文中某些句子成分，以达到译文逻辑正确、通顺流畅、重点突出等目的。

4.6.1　主语转换成宾语

这种转换技巧常用于翻译被动句，例如：

(1) As the match burns, heat and light are given off.

火柴燃烧时发出光和热。

(2) Considerable use has been made of these data.

这些资料得到了充分的利用。

(3) Much progress has been made in electrical engineering in less than a century.

译：

(4) About 40 million tons of ammonia are consumed annually in agriculture.

译：

(5) After some underground nuclear tests, large numbers of tiny aftershocks are recorded during the next few weeks.

译:

4.6.2 主语转换成谓语

以 care、need、attention、emphasis、improveme 等名词及名词化结构作主语时,常这样处理:

(1) Care should be taken at all times to protect the instrument from dust and dump.

应当始终注意保护仪器,避免沾染灰尘和受潮。

(2) There is a need for improvement in our experimental work.

译:

(3) Some improvement in efficiency can be gained at high speed by reducing viscosity and at low speed by increasing viscosity.

译:

(4) The expansion of the steam in the cylinder causes some of the steam to condense. (When the steam expands in the cylinder some of the steam condenses.)

译:

(5) The rotation of the earth on its own axis causes the change from day to night. (The earth rotates on its own axis, which causes the change from day to night.)

译:

(6) The protection of the ear to reduce sound pressure on it is necessary if the ear is to be subjected to noise at a sound pressure level more than 85 decibels continuously. (It is necessary to protect the ear to reduce sound pressure on it if the ear is to be subjected to noise at a sound pressure level more than 85 decibels continuously.)

译:

(7) The transfer of heat from one molecule to next is known as conduction.(That heat is transferred from one molecule to the next is known as conduction.)

译:

4.6.3　宾语转换成主语

英语中的常用搭配有to have a length of、to have a height of、to have a density of、to have a voltage of,如果按照字面含义翻译成汉语,则十分生硬,往往将"长度""密度""电压"作为主语;此外,词汇搭配的关系也要求成分转换,例如:

This sort of stone has a relative density of 2.7 g/cm^3.

这种石头的相对密度是2.7 g/cm^3。

Most cylinder cushions provide a relatively small velocity change.

大多数缸体缓冲器所引起的速率变化都较小。

句中动词"provide"与名词词组"velocity change"搭配很自然,但汉语中一般不说"提供速率变化",为使译文符合汉语习惯,应将"provide"意译,句子成分做相应的转换。

【练习4.4】 将下列各句翻译成汉语。

(1) An electron has about 1/1840 the mass of the proton.

译:

(2) Light beams can carry more information than radio signals because light has a much higher frequency than radio waves.

译:

4.6.4　谓语转换成主语

某些英语动词如act、behave、feature、characterize、relate、load、conduct等在汉语中往往都要转译成名词才符合汉语的表达习惯,因此,作为句子成分也就相应地转换成主语。例如:

(1) Matter is anything that occupies space and has weight.

凡占有空间并具有质量的东西都是物质。

比较:物质是占有空间并具有质量的一切东西。

为了使汉语句子结构平衡,应将句中的主语和表语对换。

(2) Neutrons act differently from protons.

译:

(3) A highly developed physical science is characterized by an extensive use of mathematics.

译:

(4) Water with salt conducts electricity very well.

译:

4.6.5 定语转换成谓语或表语

定语转换成谓语或表语往往是为了突出定语所表达的内容。例如:

(1) There is a large amount of energy wasted due to friction.

摩擦损耗了大量的能量。

(2) Many factors enter into equipment reliability.

译:

(3) The earth was formed from the same kind of materials that makes up the sun.

译:

(4) For the system V10800, for instance, an attenuation change of about ± 0.6 decibel for one repeater section can be reckoned with at the highest line frequency.

译:

4.6.6 定语与状语互相转换

由于被修饰词的词性改变,起修饰作用的定语与状语常常互相转换。此外,修饰全句的状语有时可译为定语。例如:

(1) We should have a firm grasp of the fundamentals of mechanics.

我们应牢固掌握力学的基本知识。

(2) The modern world is experiencing rapid development of science and technology.

译:

(3) The electronic computer is chiefly characterized by its accurate and rapid computations.

译:

(4) For nearly all substances the density gets smaller as the temperature is raised.

译:

4.6.7 其他转换形式

主动语态句子的某些状语可转译成主语;"there be"句型中修饰主语的of短语可从定语

转换成主语;"to have a length of "句型中的主语可以转换成定语。例如:

(1) Various substances differ widely in their magnetic characteristics.

不同物质的磁性大不相同。

(2) A word can now go by telephone from the Atlantic to the Pacific in one-twelfth (1/12) of a second.

现在电话可在1/12秒内把一句话从大西洋传到太平洋。

(3) There are three main laws of mechanics, or three laws of Newton.

力学有三大定律,即牛顿三定律。

(4) There exist many sources of energy both potential and kinetic.

势能和动能都有许多来源。

(5) Water has a variable coefficient of expansion.

水的膨胀系数是可变的。

(6) Nowadays, a typical radio transmitter has a power of 100 kilowatts so that it can broadcast information over a large area of influence.

如今,一台常见的无线电发射机的功率已达100千瓦,因此它的播送范围很大。

【练习4.5】 将下列各句翻译成汉语。

(1) The oil tank has a capacity of 50 liters.

译:

(2) Numerous materials are available to today's designers.

译:

(3) This machine tool is good in quality and small in size.

译:

(4) In any machine, input work equals output work plus work done against friction.

译:

(5) Such high-speed jet aircraft has a speed of 700 mph.

译:

(6) Extreme care must be taken to the selection of this empirical constant.

译:

(7) Heat and light can be given off by this chemical change.

译:

(8) Gas, oil and electric furnaces are commonly used for heat treating metals.

译:

(9) The same is true for graphs of total rainfall versus total storm runoff.

译:

(10) Minerals usually differ in hardness.

译:

(11) Diamonds are characterized by very great hardness.

译:

(12) Glass is much more soluble than quartz.

译:

(13) Geology is concerned with systematic study of rocks and minerals.

译:

(14) Work is done when an object is lifted.

译:

(15) The type, power and speed of the machine are seen on the name plate.

译:

(16) The expression of the relation between force, mass and acceleration is as follows.

译:

(17) Throughout the world come into use the same signs and symbols of mathematics.

译:

(18) Air travel has become faster and faster, but sometimes there is a need for traveling slowly and even for stopping in mid-air.

译:

(19) A gas has neither shape nor volume of its own but assumes those of its container.

译:

(20) In science and engineering, it is of great importance to state the laws and principles accurately.

译:

4.7　被动语态的翻译

英语中被动语态的使用极为普遍,尤其在科技英语中,这是英语区别于汉语的显著特点之一。在英语中,凡是为了强调受施者,无需提及施动者、无意点明施动者、无法说出施动者,为了上下文的衔接与连贯、出于礼貌措辞等方面的考虑不愿说出动作的执行者是谁的,往往都采用被动语态。例如:

(1) 不必说出主动者或无法说出主动者。

For a long time aluminum has been thought as an effective material for preventing metal corrosion.

(2)为了强调被动者,使其位置鲜明、突出。

Three machines can be controlled by a single operator.

(3) 为更好地联接上下文。

Vulcanized rubber was a perfect insulating material; on the rail way it was used for shock-absorbers and cushions.

化学化工专业英语中普遍使用被动语态,主要有三个方面的原因:首先,科技人员关心的是行为、活动、作用、事实,至于这些行为是谁做的,是次要的,所以,在这样的句子中就没有必要出现人称。其次,主语一般位于句首,被动句正是将"行为""活动""作用""事实"等作为主语,从而能立即引起读者对所陈述的事实的注意。最后,科技工作者为了表示客观和谦虚的态度,往往避免使用第一人称,因而尽可能使用被动语态。在翻译英语被动语态时,大量语句应译成主动句,少数句子仍可译成被动句。

4.7.1　译成主动句

(1) 当英语被动句中的主语为无生命的名词,又不出现由介词by引导的行为主体时,往往可译成主动句,原句的主语在译文中仍为主语。这种把被动语态直接译成主动语态的句子,实际是省略了"被"字的被动句。例如:

Metals are widely used in industry.

金属广泛地应用于工业。

This liquid became mixed with salt at room temperature.

这种液体在室温下和盐混合了。

（2）把原主语译成宾语，而把行为主体或相当于行为主体的介词宾语译成主语。例如：

Only a small portion of solar energy is now being used by people.

人们只能利用一小部分太阳能。

Modern scientific discoveries lead to the conclusion that energy may be created from matter and that matter in turn, may be created from energy.

近代的科学发现得出这样的结论：物质可以产生能量，能量又可以产生物质。

（3）在翻译某些被动语态时，增译适当的主语使译文通顺流畅。原文中如果没有动作的发出者，翻译时可以从逻辑出发，适当增补一些泛指的主语，如"人们""有人""大家""我们"等，并把原句的主语译成宾语。例如：

If one or more electrons are removed, the atom is said to be positively charged.

如果原子失去了一个或多个电子，我们就说这个原子带正电荷。

（4）英语中有些被动句含有地点状语或方式状语，状语中的名词常常转译成主语，而引导名词的介词常常省略不译。

Larger amounts of solvent are used in batch processed.

间接生产过程使用大量溶剂。

4.7.2 译成被动句

英语中有些被动句仍然可以译为被动句，以突出其被动意义。汉语中表达被动意义的语言手段主要包括使用"被""受""遭""让""给""为……所""加以""予以"等。

（1）在谓语前面使用"被""受""给"等字。例如：

The heated water is thus cooled as it goes through the radiator.

热水流过冷却器时就被冷却了。

Without the ozone layer, plants and animals would also be affected.

如果没有臭氧层，动植物也将受到影响。

（2）使用"为……所"句型。例如：

The atomic theory was not accepted until the last century.

原子学说直到上个世纪才为人们所接受。

（3）增加"加以""予以"等词。例如：

Other evaporation processes will be discussed briefly.

一些其他的蒸发过程将简单地加以讨论。

（4）译为汉语的"把"字句，即在原主语前加上"把"字。例如：

The mechanical energy can be changed back into electrical energy through a generator.

发电机可以把机械能再转变为电能。

4.7.3 译成无主句

英语中有许多被动句没有提到动作的发出者,这种句子常常可以译成汉语的无主句,原句中的主语译成宾语。例如:

This kind of unpleasant noise must be immediately put an end to.

必须立即终止这种令人讨厌的噪音。

Attention has been paid to the new measures to prevent corrosion.

已经注意到采取防腐新措施。

4.7.4 it作形式主语的被动句译法

由it做形式主语的被动句型,翻译时一般按主动结构译出,即将原文中的主语从句译成宾语,而把it做形式主语的主句译成一个独立语或分句。例如:

It is widely acknowledged that these materials can withstand strong stress and high temperature.

人们普遍认为,这些材料能承受强大应力和高温。

It has been proved that a material's dimensions is one of the factors influencing its ability to conduct electricity.

据证明,材料的尺寸是影响其导电能力的因素之一。

以it作形式主语的常见句型:

It is reported that... 据报道……

It is said that... 据说……

It is supposed that... 据推测……

It has been proved that... 业已证明……

It should be noted that... 应当指出……

It is assumed that... 假设,假定……

It cannot be denied that... 不可否认……

It may be said without fear of exaggeration that... 可以毫不夸张地说……

It is estimated that... 据估计……

It must be realized that... 必须了解……

It must be admitted that... 必须承认……

It must be pointed that... 必须指出……

【练习4.6】 将下列各句翻译成汉语。

(1) The molecules are held together by attractive forces.

译:

(2) To explore the moon's surface, rockets were launched again and again.

译:

(3) The metal, iron in particular, is known to be an important material in engineering.

译:

(4) When the phase boundary is covered by surfactant molecules, the surface tension is reduced.

译:

(5) Masses of oxygen is produced in air by green plants' photosynthesis to keep the living beings alive.

译:

(6) This system of units is commonly used by most mathematicians.

译:

(7) Electricity can be transmitted over a long distance.

译:

(8) The speed of the molecules is increased when they are heated.

译:

(9) Many car engines are cooled by water.

译:

(10) Several elements and compounds may be extracted directly from seawater.

译:

4.8 分译法和反译法

4.8.1 分译法

有时英语长句中主句与从句或主句与修饰语间的关系并不十分密切，翻译时可按汉语多用短句的习惯，把长句中的从句或短语转化为句子，分开来叙述。为了使语言连贯，有时还可适当增加词语。这种翻译方法称为分译法。例如：

(1) The extensive tissue penetration of ciprofloxacin combined with its enhanced antibacterial activity, enables ciprofloxacin to be used alone or in combination with an aminoglycoside or with beta-lactam antibiotics.

环丙沙星具有广泛的组织渗透性和较强的抗菌活性，这使其可单独使用或与某种氨基糖苷类抗生素或β-内酰胺类抗生素联用。

(2) Half-lives of different radioactive elements vary from as much as 900 million years for one form of uranium, to a small fraction of a second for one form of polonium.

译：

(3) The diode consists of a tungsten filament, which gives off electrons when it is heated, and a plate toward which the electrons migrate when the field is in the right direction.

译：

4.8.2 反译法

由于英国人与中国人的思维习惯不尽相同，英、汉两种语言的表达方式也有很多区别。有时英语从正面表达的思想内容，汉语需要从反面表达才更合适，反之亦然。这种翻译方法称为反译法。反译法如运用得当，往往能更确切、完整地再现全文神韵，而且译文也更通畅、规范。例如：

(1) The flowing of electricity through a wire is **not unlike** that of water through a pipe.

电流通过导线**正如**水流过管道一般。

(2) Nuclear radiation is **not harmless** to human beings and other living things.

译：

(3) To a great extent the value placed on gold and silver is due to their **lack** of reactivity.

译：

(4) The motor **refused** to start.

译:

4.9 专业术语的翻译

专业术语的翻译是化学化工专业英语翻译中的一个重点。专业术语的翻译存在着许多不同于一般英语翻译的地方。例如:element一词,一般英语译为"要素""成分",但在化学中译为"元素"、电学中译为"电极"、无线电学中译为"元件"等。再如下面这个句子:

Protophilic solvents such as liquid ammonia, amines and ketones which possess a high affinity for protons.

亲质子的溶剂是一种对质子具有很高的亲和性的物质,如液氨、胺类和酮类。

在这个句子中,affinity意思是"密切关系、姻亲关系、吸引力",但在专业英语中译为"亲和性"。

科技英语翻译必须讲究专业性,故应该学习普通英语词汇的专业化译法才能把科技英语翻译得更科学、更严谨和更完整。

在上一节讲到科技英语翻译的表达方法主要有直译和意译两种。

所谓直译,就是直接译出词汇所指的意义。科技英语翻译中大量使用的就是直译这种方法。直译包括移植译(transplant)、音译(transliteration)和形译(pictographic translation)。但有的科技英语词汇却必须采用意译。意译包括推演(deduction)、引申(extension)和解释(explanation)。

4.9.1 移植译

移植译就是按词典里所给的词义将词的各个词素的意义依次译出。翻译派生词和复合词时多采用这种方法,例如:microwave(微波)、informstion superhighway(信息高速公路)、magnetohydrodynamics(磁流体力学)。这些专业词语长而且复杂,往往是由一些基本的科技英语词素组合成的,因而,大多采用移植译法。

4.9.2 音译

专有名词(如人名、地名等)通常需要采用音译法。此外,有些词如新材料、药名、缩略词等,在汉语中没有确切的对等译词,按照意译又比较费力,就只好借助音译。例如,clone(克隆)、hacker(黑客)、nylon(尼龙)、aspirin(阿斯匹林)、radar(雷达)等,这些词都是按音译进行翻译的。也有一些词是部分音译的,如:AIDS(艾滋病)、topology(拓扑学)、Hellfire(海尔法)等。

4.9.3　形译

为了形象化,科技术语中常采用外文字母或英语单词来描述某种与技术有关的形象。翻译时可以将该外文字母照抄,改译为字形或概念、内涵相近的汉字,这种用字母或汉字来表达形状的翻译方法称为形译法。形译法可细分为以下三种:

第一种,保留原字母不译。例如:O-ring(O形环)、S-turning(S形弯道)、X-ray(X射线)、A-bedplate(A形底座)。

第二种,用汉语形象相似的词来译。例如:steel I-beam(工字钢梁)、T-bolt(T 字螺栓)、O-ring(环形圈)、L-square(直角尺)。

第三种,用能表达其形象的词来译。例如:U-bolt(马蹄螺栓)、V-belt(三角带)、T-bend(三通接头)、twist drill(麻花钻)。

4.9.4　推演

推演是根据原文本或原文词典中的意思进行概括,推演出汉语的意义。译文包含的不仅是原词的字面意义,还必须概括出词语所指事物的基本特征。例如,space shuttle,如果按照移植的方法将其译成"太空穿梭机"显然不妥,很容易引起误解。其实,这里的space指的是aerospace(航天),shuttle指往返于太空与地球之间的形状像飞机的交通工具,因此,将space shuttle推演译成"航天飞机"。这种用推演法译出的词语比用移植法更直观、易懂,因此,也更容易使人接受。推演法使用得当便能译出高质量的译文,这就要求译者不但要有较好的专业基础知识,还必须具备两种语言的良好修养。

4.9.5　引申

所谓引申就是在不脱离原文的基础上,运用延续与扩展的方法译出原文。通常的做法是:① 将具体所指引向抽象泛指,如brain具体词义是"大脑",抽象意义指"智力",brain-trust则可以引申为"智囊团"。② 将抽象泛指引向具体所指,如qualification抽象泛指"鉴定",具体可以指"通过鉴定所具备的条件",因此,data qualification可以引申为"数据限制条件"。

4.9.6　解释

若存在某个词在使用上述方法都难译好的情况下,可采用解释法,即用汉语说出英语原文的意思而不必给出汉语的对等词。如blood type可译为"血型"、blood bank可译成"血库",但blood heat却不能译成"血热",而用其他方法也很难译出其准确含义,此时可借用解释法,将其译成"人体血液正常温度"。这一方法大多用于个别初次出现而意义比较抽象、含义比较深刻的名词或术语。

【练习4.7】 将下列各句翻译成汉语。

(1) Organic multicarboxylate liands present versatile coordination modes and are very effective building blocks for the construction of coordination polymers.

译:

(2) Since 1940 the chemical industry has grown at a remarkable rate, although this has slowed significantly in recent years. The lion's share of this growth has been in the organic chemicals sector due to the development and growth of the petrochemicals area since 1950. The explosive growth in petrochemicals in the 1960s and 1970s was largely due to the enormous increase in demand for synthetic polymers such as polyethylene, polypropylene, nylon, polyesters and epoxy resins.

译:

材料阅读

Some Points of English Style

- Do not use nouns as adjectives.
- The word "this" must always be followed by a noun, so that its reference is explicit.
- Describe experimental results uniformly in the past tense.
- Use the active voice whenever possible.
- Complete all comparisons.
- Type all papers double-spaced (not single-or one-and-a-half spaced), and leave 1 space after colons, commas, and after periods at the end of sentences. Leave generous margins. (generally, 1.25" on both sides & top & bottom).

Assume that we will write all papers using the style of the American chemical society. You can get a good idea of this style from three sources:

(1) The journal. Simply look at articles in the journals and copy the organization you see there.

(2) Previous papers from the group. By looking at previous papers, you can see exactly how a paper should "look". If what you wrote looks different, it probably is not what we want.

(3) The ACS Style Guide: A Manual for Authors and Editors. (Janet S. Dodd, Editor Washington, D.C. USA 1997) . Useful detail, especially the section on references.

I also suggest you read Strunk and White, The Elements of Style (Longman: New York,

2000, 4th edition) to get a sense for English usage. Two excellent books on the design of graphs and figures are *The Visual Display of Quantitative Information* by Edward R. Tufte, Graphics Press (1983) and *Envisioning Information* also by Edward R. Tufte, Graphics Press (1990).

（摘自 http://bloy.sciencenet.cn/home.php? mod=space & uid=308 & amp; do=log & id=2078.）

第5章　化学化工基础知识阅读

5.1　分析化学

Analytical chemistry

Analytical chemistry is the science of making quantitative measurements. In practice, quantifying analysis in a complex sample becomes an exercise in problem solving. To be effective and efficient, analyzing samples requires expertise in:

- The chemistry that can occur in a sample.
- Analysis and sample handling methods for a wide variety of problems (the tools-of-the-trade).
- Proper data analysis and record keeping.

To meet these needs, Analytical Chemistry courses usually emphasize equilibrium, spectroscopic and electrochemical analysis, separations, and statistics.

Analytical chemistry requires a broad background knowledge of chemical and physical concepts. With a fundamental understanding of analytical methods, a scientist faced with a difficult analytical problem can apply the most appropriate technique(s). A fundamental understanding also makes it easier to identify when a particular problem cannot be solved by traditional methods, and gives an analyst the knowledge that is needed to develop creative approaches or new analytical methods.

【练习5.1】 补全下列分析化学仪器名称。

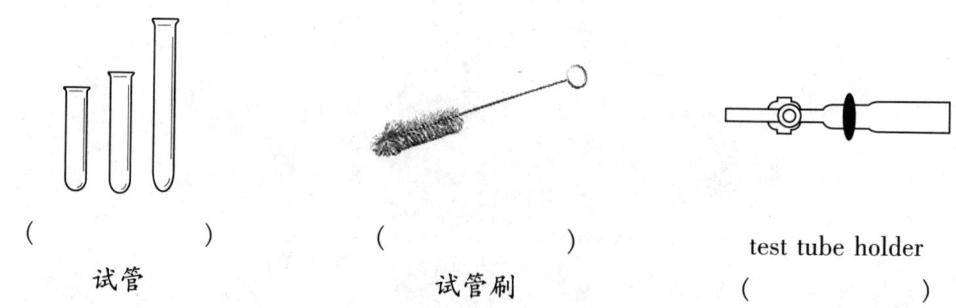

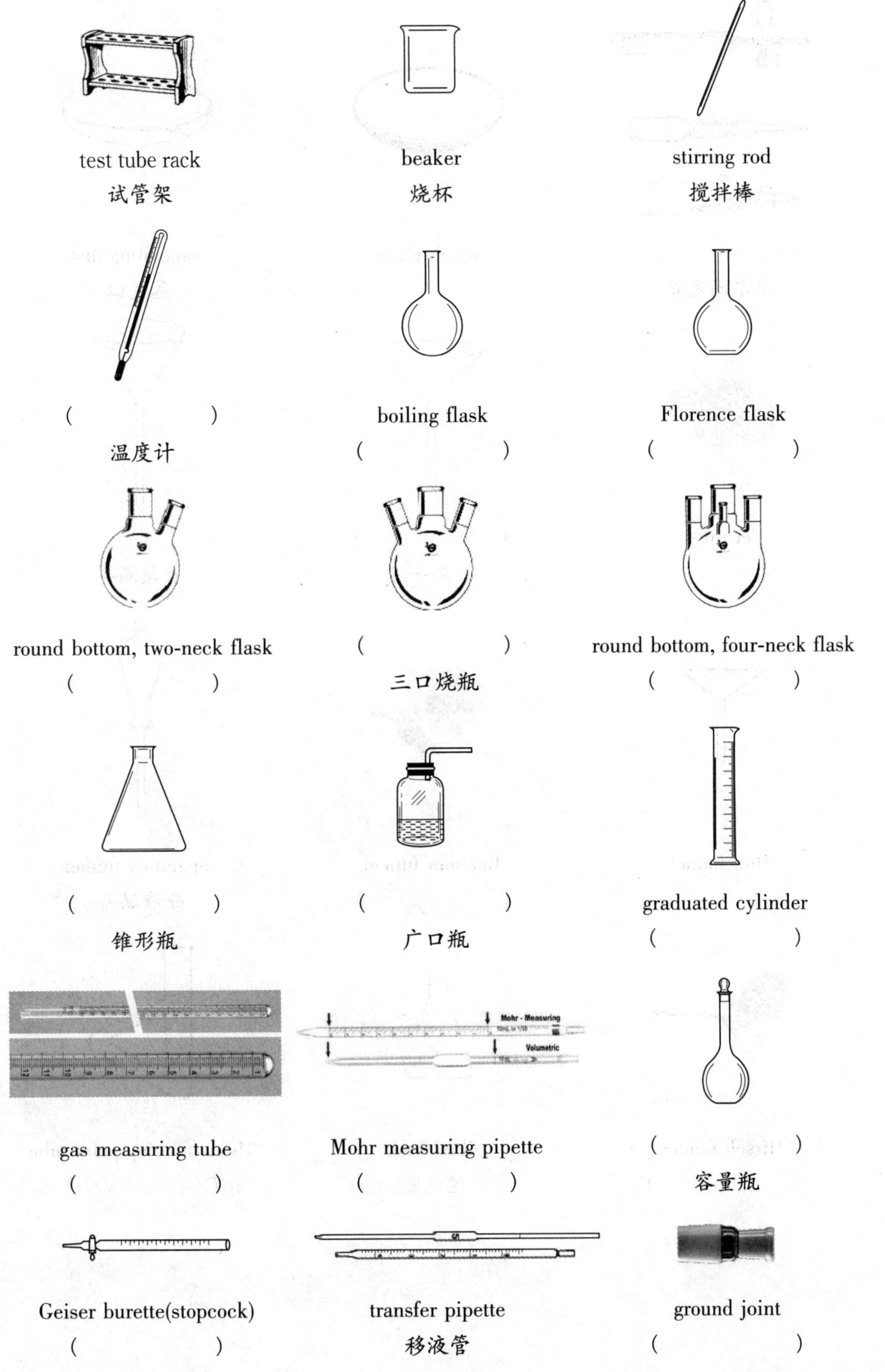

test tube rack
试管架

beaker
烧杯

stirring rod
搅拌棒

(　　　　)
温度计

boiling flask
(　　　　)

Florence flask
(　　　　)

round bottom, two-neck flask
(　　　　)

(　　　　)
三口烧瓶

round bottom, four-neck flask
(　　　　)

(　　　　)
锥形瓶

(　　　　)
广口瓶

graduated cylinder
(　　　　)

gas measuring tube
(　　　　)

Mohr measuring pipette
(　　　　)

(　　　　)
容量瓶

Geiser burette(stopcock)
(　　　　)

transfer pipette
移液管

ground joint
(　　　　)

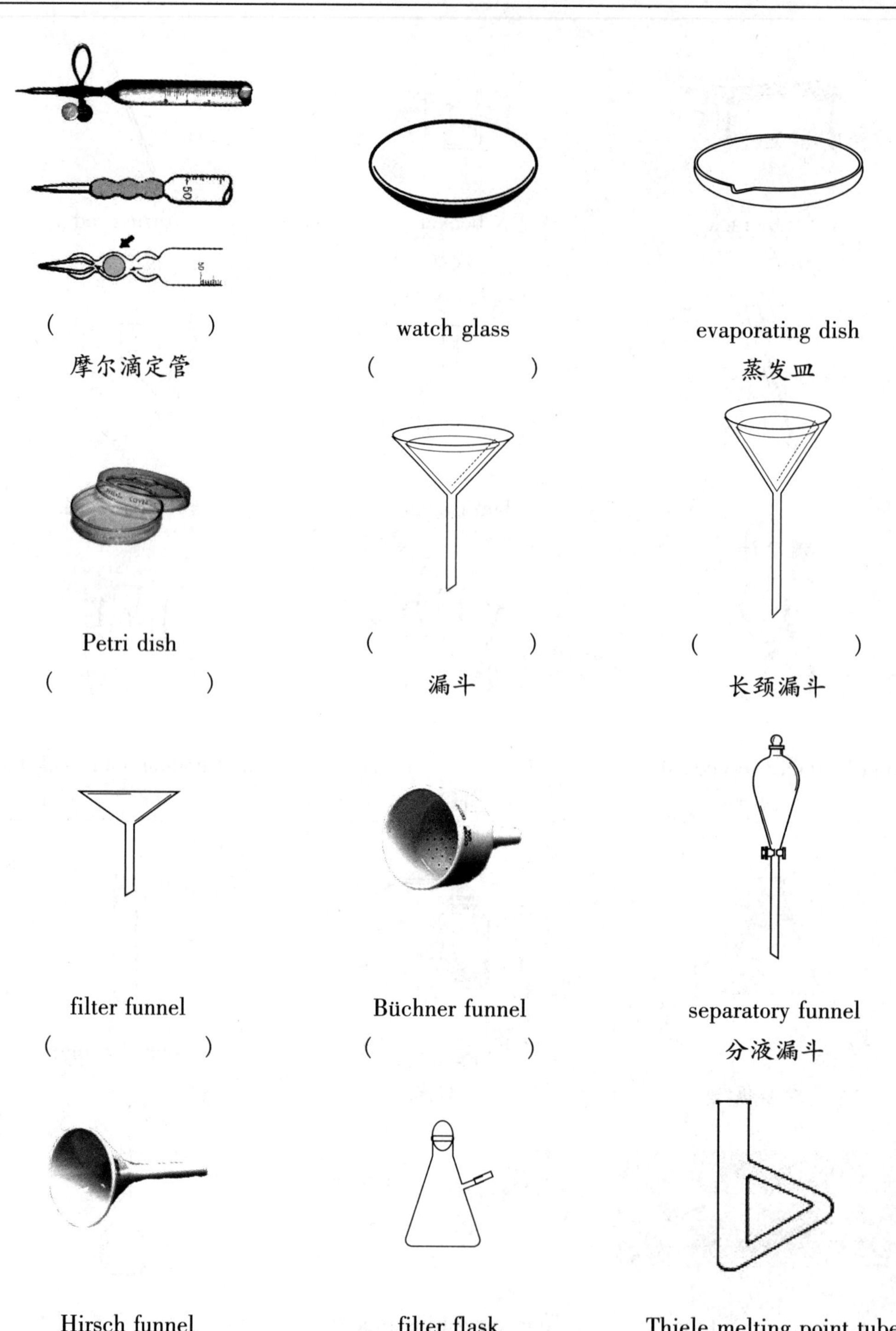
(　　　　)
摩尔滴定管
watch glass
(　　　　)
evaporating dish
蒸发皿
Petri dish
(　　　　)
(　　　　)
漏斗
(　　　　)
长颈漏斗
filter funnel
(　　　　)
Büchner funnel
(　　　　)
separatory funnel
分液漏斗
Hirsch funnel
(　　　　)
filter flask
过滤瓶
Thiele melting point tube
(　　　　)

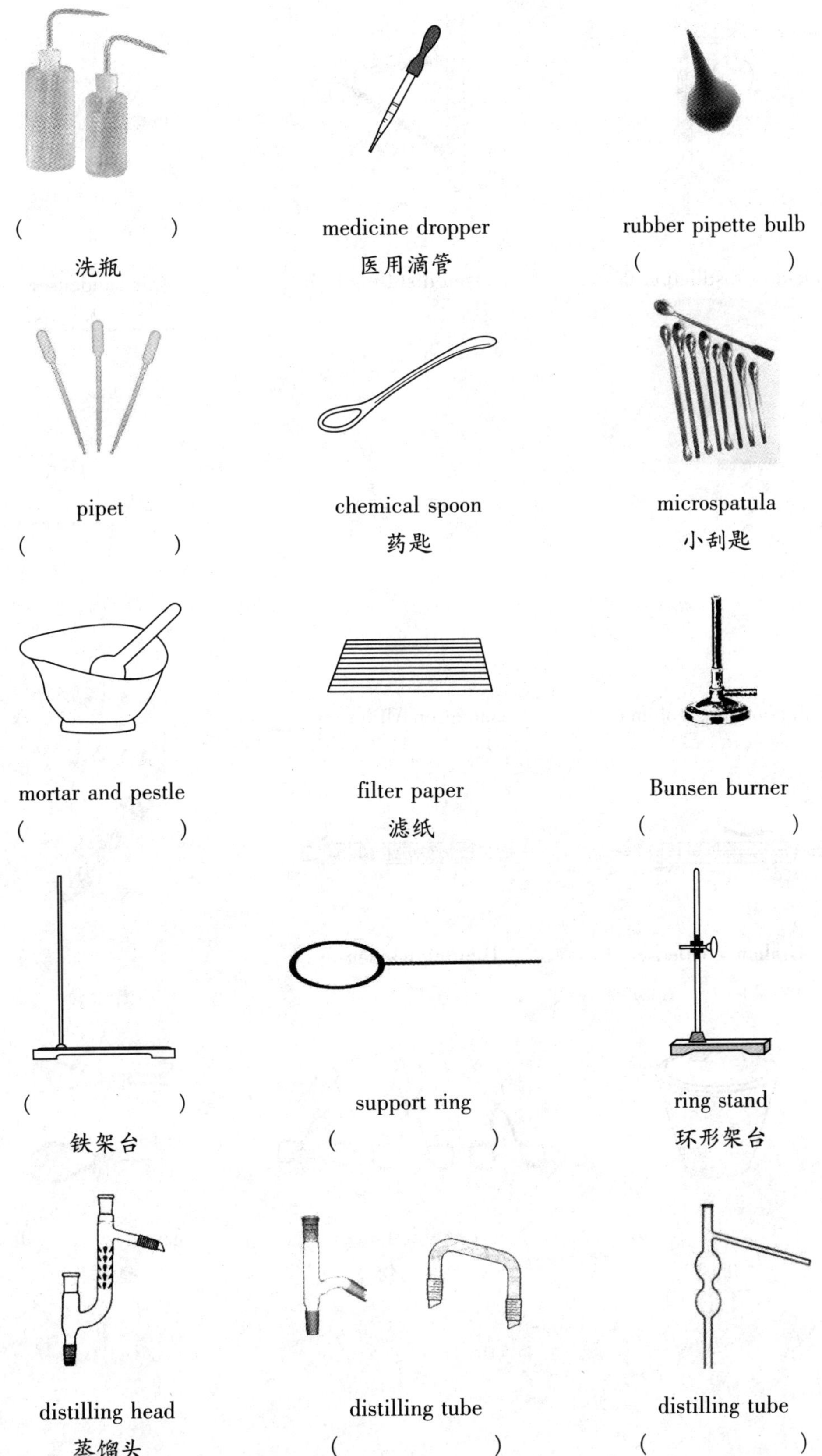
(　　　　　)
洗瓶
medicine dropper
医用滴管
rubber pipette bulb
(　　　　　)
pipet
(　　　　　)
chemical spoon
药匙
microspatula
小刮匙
mortar and pestle
(　　　　　)
filter paper
滤纸
Bunsen burner
(　　　　　)
(　　　　　)
铁架台
support ring
(　　　　　)
ring stand
环形架台
distilling head
蒸馏头
distilling tube
(　　　　　)
distilling tube
(　　　　　)

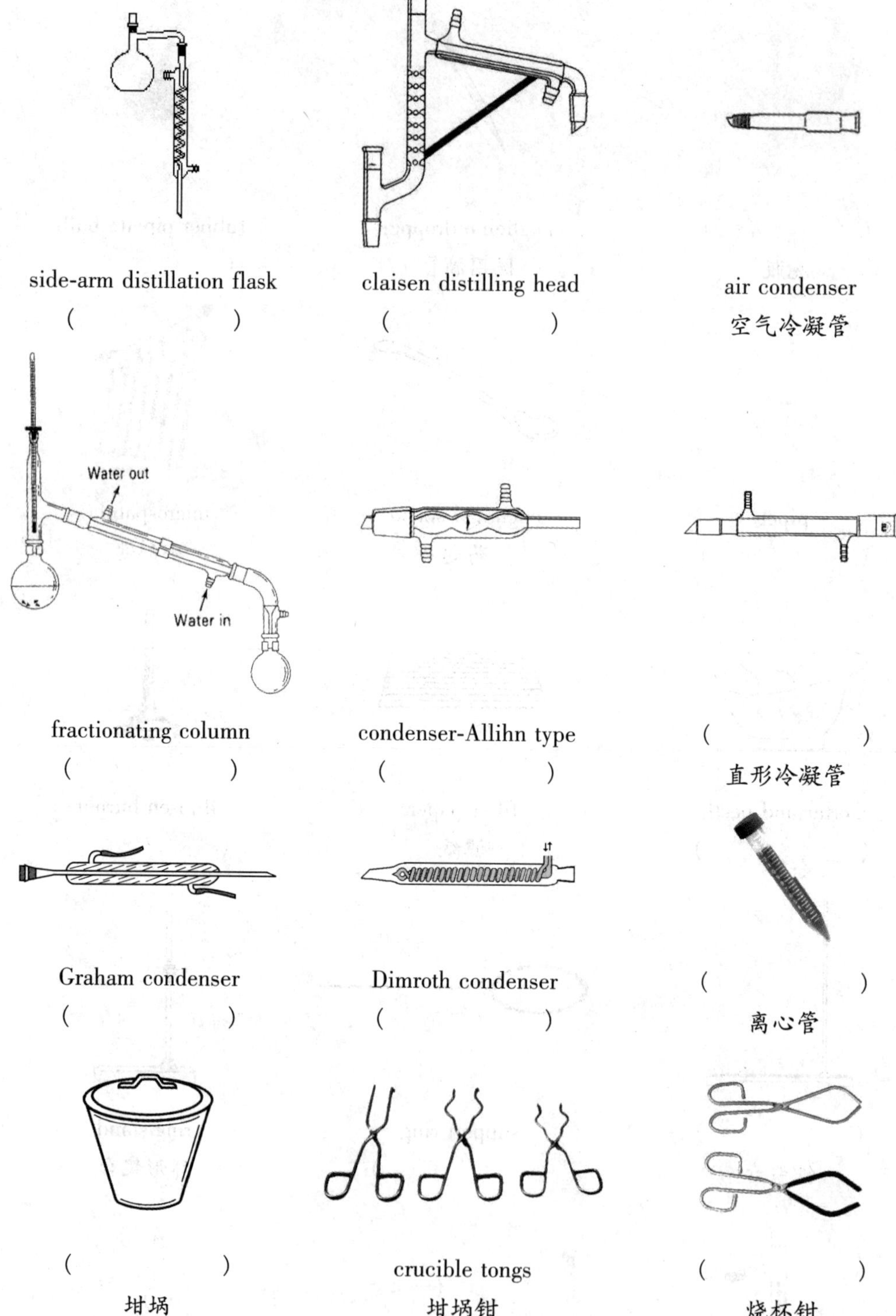

side-arm distillation flask
(　　　　　　)

claisen distilling head
(　　　　　　)

air condenser
空气冷凝管

fractionating column
(　　　　　　)

condenser-Allihn type
(　　　　　　)

(　　　　　　)
直形冷凝管

Graham condenser
(　　　　　　)

Dimroth condenser
(　　　　　　)

(　　　　　　)
离心管

(　　　　　　)
坩埚

crucible tongs
坩埚钳

(　　　　　　)
烧杯钳

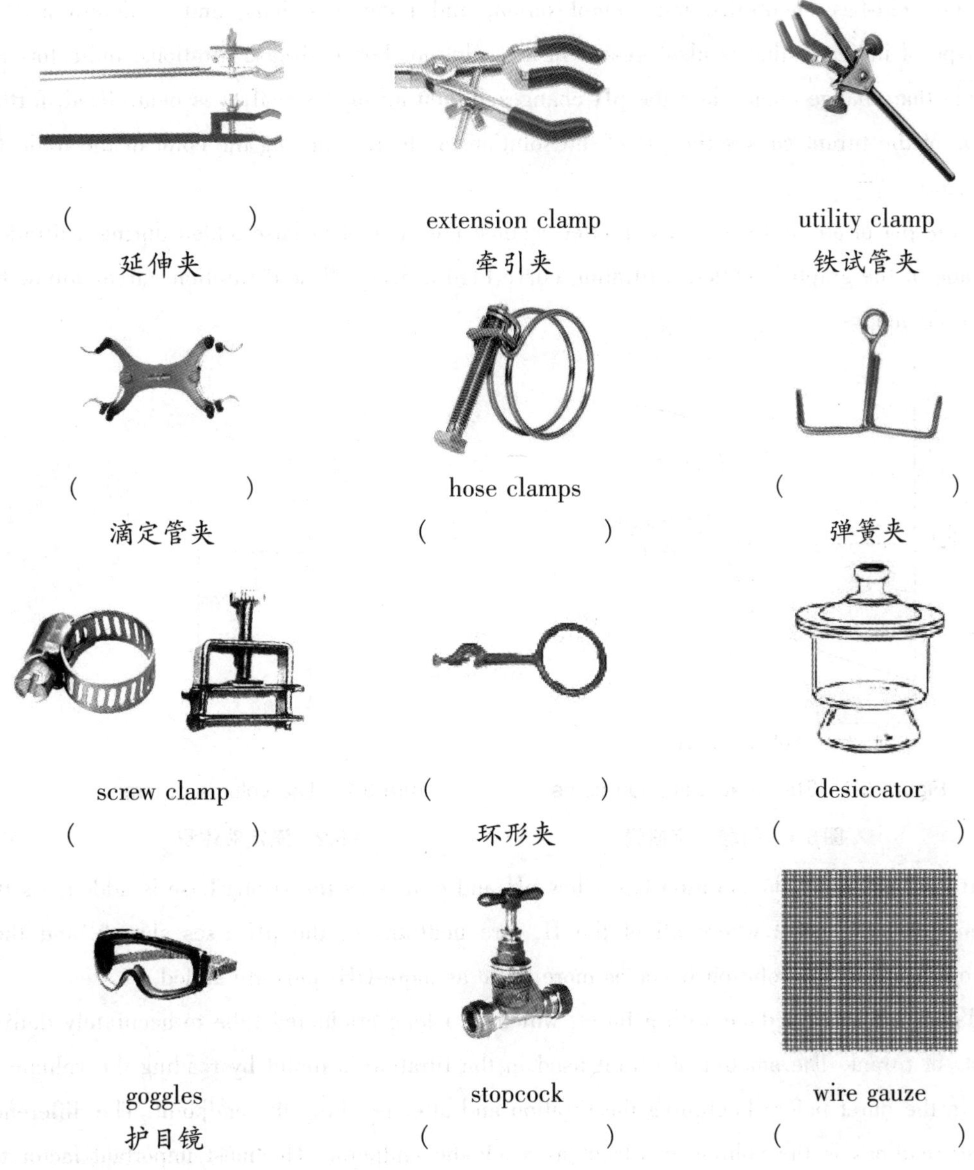

材料阅读

Titration

Titration is the quantitative measurement of an analyte in solution by completely reacting it with a reagent solution. The reagent is called the titrant and must either be prepared from a primary standard or be standardized versus a primary standard to know its exact concentration.

The point at which all of the analyte is consumed is the equivalence point. The number of moles of analyte is calculated from the volume of reagent that is required to react with all of the analyte, the titrant concentration, and the reaction stoichiometry.

The equivalence point is often determined by visual indicators are available for titrations based on acid-base neutralization, complexation, and redox reactions, and is determined by some type of indicator that is also present in the solution. For acid-base titrations, indicators are available that change color when the pH changes. When all of the analyte is neutralized, further addition of the titrant causes the pH of the solution to change causing the color of the indicator to change.

If the pH of an acid solution is plotted against the amount of base added during a titration, the shape of the graph is called a titration curve(Figure 5.1). All acid titration curves follow the same basic shapes.

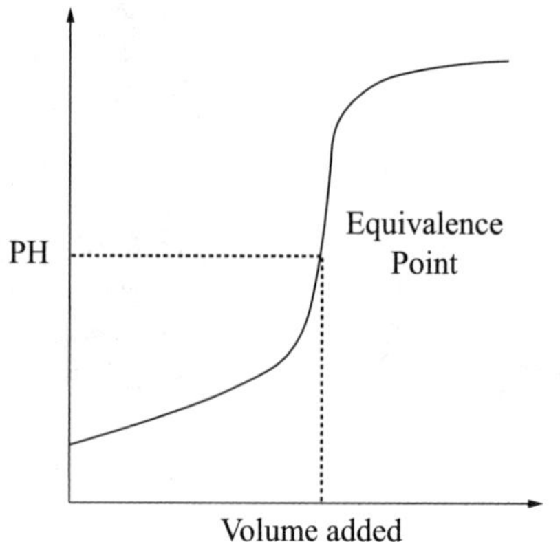

Figure 5.1 Strong acid titration curve

图5.1 强酸滴定曲线

Figure 5.2 The volume of titrant

图5.2 滴定液体积

At the beginning, the solution has a low pH and climbs as the strong base is added. As the solution nears the point where all of the H^+ are neutralized, the pH rises sharply and then levels out again as the solution becomes more basic as more OH^- ions are added.

Manual titration is done with a buret, which is a long graduated tube to accurately deliver amounts of titrant. The amount of titrant used in the titration is found by reading the volume of titrant in the buret before beginning the titration and after reaching the endpoint. The difference in these readings is the volume of titrant to reach the endpoint. The most important factor for making accurate titrations is to read the buret volumes reproducibly. The figure 5.2 shows how to do so by using the bottom of the meniscus to read the reagent volume in the buret.

The end point can be determined by an indicator as described above or by an instrumental method. The most common instrumental detection method is potentiometric detection. The equivalence point of an acid-base titration can be detected with a pH electrode. Titrations, such as complexation or precipitation, involving other ions can use an ion-selective electrode (ISE). UV-VIS absorption spectroscopy is also common, especially for complexometric titrations where a subtle color change occurs.

For repetitive situations, autotitrators with microprocessors are available that deliver the titrant, stop at the endpoint, and calculate the concentration of the analyte. The endpoint is usually detected by some type of electrochemical measurement. Some examples of titrations for which autotitrators are available include:

• Acid or base determination by pH measurement with potentiometric detection.

• Determination of water by Karl Fischer reagent (I_2 and SO_2 in methyl alcohol and pyridine) with coulometric detection.

• Determination of Cl in aqueous solution with phenylarsene oxide using amperometric detection.

（摘自 http://www.cartage.org.lb/en/themes/sciences/Chemistry/Analyticalchemistry/mainpage.htm.）

分析化学常用词汇

absorbance 吸光度
absorbent 吸附剂
absorption curve 吸收曲线
absorption peak 吸收峰
absorptivity 吸收系数
accident error 偶然误差
accuracy 准确度
acid-base titration 酸碱滴定
acidic effective coefficient 酸效应系数
acidic effective curve 酸效应曲线
acidity constant 酸度常数
activity 活度
activity coefficient 活度系数
adsorption 吸附
adsorption indicator 吸附指示剂
affinity 亲和力
aging 陈化
amorphous precipitate 无定形沉淀
amphiprotic solvent 两性溶剂
amphoteric substance 两性物质
amplification reaction 放大反应
analytical balance 分析天平
analytical chemistry 分析化学
analytical concentration 分析浓度
analytical reagent (AR) 分析试剂
apparent formation constant 表观形成常数
aqueous phase 水相
argentimetry 银量法
ashing 灰化
atomic spectrum 原子光谱
autoprotolysis constant 质子自递常数
auxochrome group 助色团
back extraction 反萃取
band spectrum 带状光谱
bandwidth 带宽
bathochromic shift 红移
blank 空白
blocking of indicator 指示剂的封闭(现象)
bromometry 溴量法
buffer capacity 缓冲容量
buffer solution 缓冲溶液
burette 滴定管
calconcarboxylic acid 钙指示剂
calibrated curve 校准曲线
calibration 校准
catalyzed reaction 催化反应
cerimetry 铈量法
charge balance 电荷平衡
chelate 螯合物
chelate extraction 螯合物萃取
chemical analysis 化学分析
chemical factor 化学因素

chemically pure 化学纯
chromatography 色谱法
chromophoric group 发色团
coefficient of variation 变异系数
color reagent 显色剂
color transition point 颜色转变点
colorimeter 比色计
colorimetry 比色法
column chromatography 柱色谱
complementary color 互补色
complex 络合物
complexation 络合反应
complexometry complexometric titration 络合滴定法
complexone 氨羧络合剂
concentration constant 浓度常数
conditional extraction constant 条件萃取常数
conditional formation coefficient 条件形成常数
conditional potential 条件电位
conditional solubility product 条件溶度积
confidence interval 置信区间
confidence level 置信水平
conjugate acid-base pair 共轭酸碱对
constant weight 恒量
contamination 沾污
continuous extraction 连续萃取
continuous spectrum 连续光谱
coprecipitation 共沉淀
correction 校正
correlation coefficient 相关系数
crucible 坩埚
crystalline precipitate 晶形沉淀
cumulative constant 累积常数
curdy precipitate 凝乳状沉淀
degree of freedom 自由度
demasking 解蔽
derivative spectrum 导数光谱
desiccant; drying agent 干燥剂
desiccator 保干器
determinate error 可测误差
deuterium lamp 氘灯
deviation 偏差
deviation average 平均偏差
dibasic acid 二元酸
dichloro fluorescein 二氯荧光黄
dichromate titration 重铬酸钾法
dielectric constant 介电常数
differential spectrophotometry 示差光度法
differentiating effect 区分效应
dispersion 色散
dissociation constant 离解常数
distillation 蒸馏
distribution coefficient 分配系数
distribution diagram 分布图
distribution ratio 分配比
double beam spectrophotometer 双光束分光光度计
dual-pan balance 双盘天平
dual-wavelength spectrophotometry 双波长分光光度法
electronic balance 电子天平
electrophoresis 电泳
eluent 淋洗剂
end point 终点
end point error 终点误差
enrichment 富集
eosin 曙红
equilibrium concentration 平衡浓度
equimolar series method 等摩尔系列法
Erelenmeyer flask 锥形瓶
eriochrome black T (EBT) 铬黑 T
error 误差
ethylenediamine tetraacetic acid (EDTA) 乙

二胺四乙酸
evaporation dish　蒸发皿
exchange capacity　交换容量
extent of crosslinking　交联度
extraction constant　萃取常数
extraction rate　萃取率
extraction spectrphotometric method　萃取光度法
Fajans method　法扬斯法
ferroin　邻二氮菲亚铁离子
filter　漏斗
filter　滤光片
filter paper　滤纸
filtration　过滤
fluex　溶剂
fluorescein　荧光黄
flusion　熔融
formation constant　形成常数
frequency　频率
frequency density　频率密度
frequency distribution　频率分布
gas chromatography (GC)　气相色谱
grating　光栅
gravimetric factor　重量因素
gravimetry　重量分析
guarantee reagent (GR)　保证试剂
high performance liquid chromatography (HPLC)高效液相色谱
histogram　直方图
homogeneous precipitation　均相沉淀
hydrogen lamp　氢灯
hypochromic shift　紫移
ignition　灼烧
indicator　指示剂
induced reaction　诱导反应
inert solvent　惰性溶剂
instability constant　不稳定常数
instrumental analysis　仪器分析
intrinsic acidity　固有酸度
intrinsic basicity　固有碱度
intrinsic solubility　固有溶解度
iodimetry　碘滴定法
iodine-tungsten lamp　碘钨灯
ion association extraction　离子缔合物萃取
ion chromatography (IC)　离子色谱
ion exchange　离子交换
ion exchange resin　离子交换树脂
ionic strength　离子强度
isoabsorptive point　等吸收点
Karl Fisher titration　卡尔·费歇尔滴定法
Kjeldahl determination　凯氏定氮法
Lambert-Beer law　朗伯-比尔定律
leveling effect　拉平效应
ligand　配位体
light source　光源
line spectrum　线状光谱
linear regression　线性回归
liquid chromatography (LC)　液相色谱
macro analysis　常量分析
masking　掩蔽
masking index　掩蔽指数
mass balance　物料平衡
matallochromic indicator　金属指示剂
maximum absorption　最大吸收
average　平均值
measured value　测量值
measuring cylinder　量筒
measuring pipette　吸量管
median　中位数
mercurimetry　汞量法
mercury lamp　汞灯
mesh　(筛)目
methyl orange (MO)　甲基橙
methyl red (MR)　甲基红

micro analysis 微量分析
mixed constant 混合常数
mixed crystal 混晶
mixed indicator 混合指示剂
mobile phase 流动相
Mohr method 莫尔法
molar absorptivity 摩尔吸收系数
mole ratio method 摩尔比法
molecular spectrum 分子光谱
monoacid 一元酸
monochromatic color 单色光
monochromator 单色器
neutral solvent 中性溶剂
neutralization 中和
non-aqueous titration 非水滴定
normal distribution 正态分布
occlusion 闭塞
organic phase 有机相
ossification of indicator 指示剂的僵化
outlier 离群值
oven 烘箱
paper chromatography(PC) 纸色谱
parallel determination 平行测定
path lenth 光程
permanganate titration 高锰酸钾滴定法
phase ratio 相比
phenolphthalein (PP) 酚酞
photocell 光电池
photoelectric colorimeter 光电比色计
photometric titration 光度滴定法
photomultiplier 光电倍增管
phototube 光电管
pipette 移液管
polar solvent 极性溶剂
polyprotic acid 多元酸
population 总体
postprecipitation 后沉淀
precipitant 沉淀剂
precipitation form 沉淀形式
precipitation titration 沉淀滴定法
precision 精密度
preconcentration 预富集
predominance-area diagram 优势区域图
primary standard 基准物质
prism 棱镜
probability 概率
proton 质子
proton condition 质子条件
protonation 质子化
protonation constant 质子化常数
purity 纯度
qualitative analysis 定性分析
quantitative analysis 定量分析
quartering 四分法
random error 随机误差
range 全距(极差)
reagent blank 试剂空白
reagent bottle 试剂瓶
recording spectrophotometer 自动记录式分光光度计
recovery 回收率
redox indicator 氧化还原指示剂
redox titration 氧化还原滴定法
referee analysis 仲裁分析
reference level 参考水平
reference material (RM) 标准物质
reference solution 参比溶液
relative error 相对误差
resolution 分辨力
rider 游码
routine analysis 常规分析
sample 样本,样品
sampling 取样
self indicator 自身指示剂

semimicro analysis　半微量分析
separation　分离
separation factor　分离系数
side reaction coefficient　副反应系数
significance test　显著性检验
significant figure　有效数字
simultaneous determination of multiponents　多组分同时测定
single beam spectrophotometer　单光束分光光度计
single-pan balance　单盘天平
slit　狭缝
sodium diphenylamine sulfonate　二苯胺磺酸钠
solubility product　溶度积
solvent extraction　溶剂萃取
species　物种
specific extinction coefficient　比消光系数
spectral analysis　光谱分析
spectrophotometer　分光光度计
spectrophotometry　分光光度法
stability constant　稳定常数
standard curve　标准曲线
standard deviation　标准偏差
standard potential　标准电位
standard series method　标准系列法
standard solution　标准溶液
standardization　标定
starch　淀粉
stationary phase　固定相
steam bath　蒸气浴
stepwise stability constant　逐级稳定常数
stoichiometric point　化学计量点
structure analysis　结构分析
supersaturation　过饱和
systematic error　系统误差
test solution　试液
thermodynamic constant　热力学常数
thin layer chromatography (TLC)　薄层色谱
titrand　被滴物
titrant　滴定剂
titration　滴定
titration constant　滴定常数
titration curve　滴定曲线
titration error　滴定误差
titration index　滴定指数
titration jump　滴定突跃
titrimetry　滴定分析
trace analysis　痕量分析
transition interval　变色间隔
transmittance　透射比
triacid　三元酸
true value　真值
tungsten lamp　钨灯
ultratrace analysis　超痕量分析
UV-VIS spectrophotometry　紫外-可见分光光度法
volatilization　挥发
Volhard method　福尔哈德法
volumetric flask　容量瓶
volumetry　容量分析
wash bottle　洗瓶
washings　洗液
water bath　水浴
weighing bottle　称量瓶
weighting form　称量形
weights　砝码
xylenol orange (XO)　二甲酚橙
working curve　工作曲线
zero level　零水平

5.2 物理化学

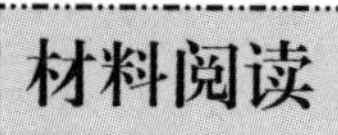

Physical Chemistry

Physical chemistry looks at chemistry from a different angle to inorganic and organic chemistry. Whereas these branches study the reactions of different groups of elements and compounds, physical chemistry uses mathematics and physics to determine laws which describe the way chemical reaction happen.

1. Energy, Structure and Rate

So physical chemistry deals with Thermodynamics or the energy levels and changes involved in chemicals and chemical reactions. It also deals with the Quantum Mechanics of chemicals. This is the atomic and molecular structure of the chemicals. Finally it deals with kinetics or the rates of reactions and how they are changed.

2. Obscure but Fundamental

It could be argued that this is the least interesting of the branches of chemistry. It involves lots of mathematical formulae and obscure theories involving Greek letters. Of course these concepts are also vital for many important processes in nature and in many of the equipment we take for granted.

3. Energy Changes

Thermodynamics involve changes in energy. Changes in heat energy cause changes in temperature, which is foundational to cooking, refrigeration and transport. Chemical reactions which produce heat are exothermic, such as in burning fuel or an exploding bomb. This usually means that chemical bonds have been broken. Chemical reactions which take heat in are called endothermic and usually mean that chemical bonds have been formed.

4. Atomic Structure

Quantum mechanics deals with the structure of the very foundational units of all matter, atoms. So the arrangement of atoms in different solids such as steel, concrete and ice are determined by this discipline. But this study also led to the discovery of the structure of the atomic nucleus leading to both the atomic bomb and nuclear energy.

5. Speed of Reaction

Kinetics is about speed of reaction. We depend on reaction rates in so many ways in our modern lives. If reactions in the body did not involve enzymes which control the rate of the reactions, we would not be able to live due to the reactions happening too slowly. The different cooking methods demonstrate the way different conditions affect the rate of reaction; to bake a potato takes hours in a conventional oven, but only a matter of minutes in a microwave oven.

6. Catalysts

The study of the way rates of reaction can be changed by catalysts like enzymes or precious metals such as platinum or rhodium is important. Reverse catalysts or inhibitors are important in the preservation of foodstuffs and have been greatly developed.

7. Collision Theory

The way conditions affect reaction rates is all about collisions(Figure 5.3). For example as temperature increases the particles move around faster and so are more likely to collide and react quicker. Higher concentration also causes more collisions. This is an important part of the study of kinetics.

Physical chemistry is a less glamorous but foundational part of scientific study.

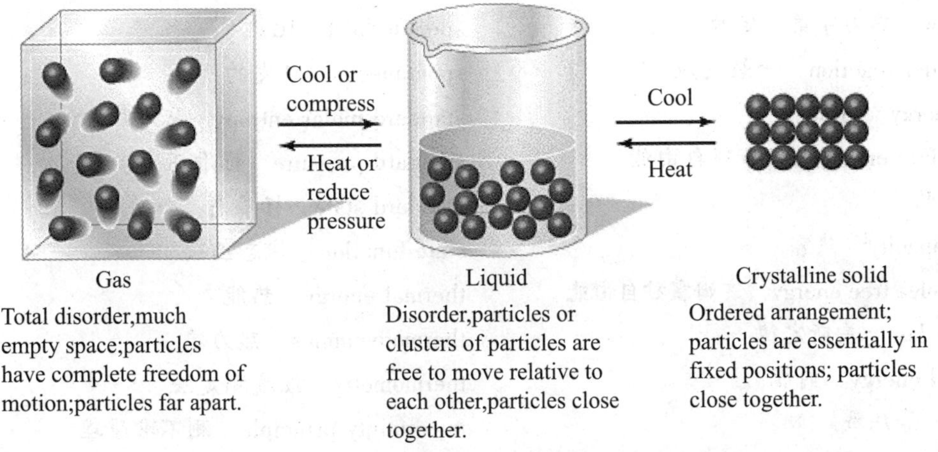

Figure 5.3　The schematic diagram of collision theory

图5.3　碰撞理论示意图

（摘自　http://molecular-chemistry.suite101.com/article.cfm/what is physical chemistry.）

物理化学常用词汇

activation energy　活化能

adiabat　绝热线

adiabatic　绝热的

amplitude　振幅，波峰

Boltzmann constant　玻尔兹曼常数

Boltzmann equation　玻尔兹曼方程

calorie　卡路里

calorimeter　热量计

collision theory　碰撞理论
empirical temperature　假定温度
endothermic　吸热(性)的
enthalpy　焓
enthalpy of combustion　燃烧焓
enthalpy of fusion　熔化热
enthalpy of hydration　水合热
enthalpy of reaction　反应热
enthalpy of sublimation　升华热
enthalpy of vaporization　汽化热
entropy　熵
exothermic　发热的
enthalpy change　焓变
enthalpy of formation　生成焓
enthalpy of reaction　标准反应焓
entropy of reaction　标准反应熵
first law　热力学第一定律
first order reaction　一级反应
free energy　自由能
Gibbs free energy　吉布斯自由能
heat　热
heat capacity　热容
Helmholez free energy　亥姆霍兹自由能
Hess's law　盖斯定律
internal energy　内能
isobar　等压线
isobaric　等压的
isochore　等容线,等体积线
isochoric　等容的
isotherm　等温线
isothermal　等温的
joule　焦耳
kinetic energy　动能
latent heat　潜热
momentum　动量,要素
Planck constant　普朗克常数
potential energy　势能
quantum　量子,量子论
quantum mechanics　量子力学
quantum number　量子数
rate law　速率定律
second law of thermodynamics　热力学第二定律
second order reaction　二级反应
specific heat　比热
spontaneous　自发的
standard molar entropy　标准摩尔熵
standard pressure　标准压力
standard state　标准态
state function　状态函数
thermal energy　热能
thermodynamics　热力学
thermometry　温度测定法
uncertainty principle　测不准原理
zero order reaction　零级反应
zero point energy　零点能

5.3　化工操作单元

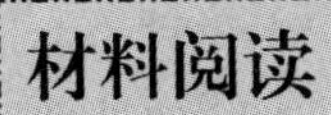

Classification of Unit Operations

1. Fluid Flow

This concerns the principles that determine the flow or transportation of any fluid from one point to another.

2. Heat Transfer

This unit operation deals with the principles that govern accumulation and transfer of heat and energy from one place to another.

3. Evaporation

This is a special case of heat transfer, which deals with the evaporation of a volatile solvent such as water from a nonvolatile solute such as salt or any other material in solution.

4. Drying

In this operation volatile liquids, usually water, are removed from solid materials.

5. Distillation

This is an operation whereby components of a liquid mixture are separated by boiling because of their differences in vapor pressure.

6. Absorption

In this process a component is removed from a gas stream by treatment with a liquid.

7. Membrane separation

This process involves the diffusion of a solute from a liquid or gas through a semipermeable membrane barrier to another fluid.

8. Liquid-Liquid Extraction

In this case a solute in a liquid solution is removed by contacting with another liquid solvent which is relatively immiscible with the solution.

9. Liquid-Solid Leaching

This involves treating a finely divided solid with a liquid that dissolves out and removes a solute contained in the solid.

10. Crystallization

This concerns the removal of a solute such as a salt from a solution by precipitating the solute from the solution.

11. Mechanical Physical Separation

These involves separation of solids, liquids, or gases by mechanical means, such as filtration, settling, and size reduction, which are often classified as separate unit operations.

Fractional distillation is the separation of a mixture of compounds by their boiling point, by heating to high enough temperatures.

（摘自 邵荣，许伟，吕慧华．新编化学化工专业英语[M]．2版．上海：华东理工大学出版社，2017.）

化工操作单元常用词汇

absorber 吸收塔
absorption 吸收
accessory 附属设备
acidity 酸度
adsorber 吸附(收)器
adsorption 吸附
air lift mud pump 气力提升泵
alkalinity 碱度
anhydrous 无水的
apparatus 仪器设备
appearance 外观
axial flow pump 轴流泵
baffle 挡板
ballmill 球磨机
batch process 间歇过程
bending moment 弯曲度
boiler 锅炉
boiling point 沸点
brine 盐水
bubble cap 泡罩
burner 烧嘴
calciner 煅烧炉
caloination 煅烧
carbonization 碳化
catalyst 催化剂
catalytical reaction 催化反应
centrifugal compressor 离心式压缩机
centrifugal pump 离心泵
centrifuger 离心机
chimney 升气管
circulation 循环
clarifier 澄清器
classifier 分级器
colorless 无色的
combustible gas 可燃气体
combustion 燃烧
combustion heat 燃烧热
composition 成分
compound 化合物
compression 压缩
concentration 浓度
condenser 冷凝器
continuous process 连续过程
control 控制
cooler 冷却器
cooling tower 冷却塔

crusher 破碎机
crushing 破碎
crystal 结晶
crystallization 结晶化
crystallizer 结晶器
decantation 倾析
decomposition 分解
deep well pump 深井泵
deflection 挠度
density 密度
desorption 解吸
dissociation 离解
distillation 蒸馏
double action reciprocating pump 双作用往复泵
downcomer 降液管
draw nozzle 抽出口
drying 干燥
ejector 喷射器
electric heater 电加热器
electric precipitator 电除尘器
electrolytic cell 电解槽
electrolyzer 电解剂
emulsion 乳化物
evaporation rate 蒸发率
evaporator 蒸发器
explosive atmosphere(environment) 爆炸性环境
explosive concentration limit 爆炸极限
explosive mixture 爆炸混合物
extract 萃出物
extraction 萃取
extraction column 萃取塔
extractor 萃取器
feeder 加料器
filter 过滤器
filter press 压滤器
filtration 过滤
fixed bed reactor 固定床反应器
flake 片状粉末
flame arrester 阻火器
float valve 浮阀
flocculants 絮凝剂
flocculation 絮凝
flocculator 浮洗器
flotation 浮选
fluidized bed reactor 流化床反应器
force and moment 力和力矩
fractionating tower 精馏塔
furnace 工业炉
gas 气体
gasification 气化
gear pump 齿轮泵
granule 颗粒
grinder 研磨机
hand (wobble) pump 手摇泵
heat exchanger 换热器
heater 加热器
helical screw compressor 螺杆式压缩机
hermetically sealed magnetic drive pump 封闭式电磁泵
high pressure vessel 高压容器
hopper 料斗
hydrogenation 氢化
hydrolysis 水解
ignition 引燃
incinerator 焚化炉
inclined rotor pump 斜转子泵
inflammable liquid 易燃液体
inline pump 管道泵
jet pump 喷射泵
leaching 淋洗
lean liquor 贫液
line mixers 管道混合器

liquid　液体
liquid distributor　液体分配(布)器
master liquor　母液
melting point　熔点
metering pump　计量泵
mixing tanks　混合槽
mixture　混合物
modulate　调节
mutche filter　抽滤器
mutiple stages compressor　多级压缩机
nash compressor　水环式压缩机
neutralization　中和
noise pollution　噪声
off-gas　废气
operation　操作
organic material　有机物
oxidation　氧化
oxidizing agent　氧化物
package　包装
package unit　成套设备
packed column　填料塔
percentage　百分比
plate column　板式塔
powder　粉末
precipitation　沉淀
pressure filter　压滤机
prill tower　造粒塔
process description　工艺叙述
process feature　工艺特点
pulverizer　粉碎机
pump　泵
raffinate　萃余物
reaction　反应
reactivation　再活化
reactor　反应器
reboiler　再沸器
reciprocating compressor　往复式压缩机
reducing agent　还原剂
reduction　还原
refrigeration　冷冻
regeneration　再生
regenerator　再生塔
relative humidity(RH)　相对湿度
rich liquor　富液
rotary compressor　旋转式压缩机
rotary kiln　回转窑
rotary pump　转子泵
screen　筛子
screw (spiral) pump　螺杆泵
scrubber　洗涤器
scrubber　洗涤塔
sedimentation　沉降
self-ignition temperature　自燃点
side reaction　副反应
sieve plate　筛板
siphon　虹吸管
solid　固体
solidificalion point　凝固点
solidification　凝固,结晶
solubility　溶解度
solution　溶解
spacer　隔叶块
specific gravity　比重
spiral plate heat exchanger　螺旋板式换热器
sprayer　喷头
static mixers　静态混合器
steam turbine　汽轮机
stress-strain diagram　应力-应变曲线
stripper　汽提塔
sublimation　升华
submersible pump　潜水泵
sump　废液池
support plate　(填料)支持板

suspension 悬浮
synthetics 合成
thickener 增稠器
top (bottom) tray 顶(底)层塔盘
tower accessories 塔附件
translucent 半透明的
transparent 透明的
tubular heat exchanger 管热式换热器
turbine pump 涡轮泵
vacuum 真空
vacuum pump 真空泵
viscosity 黏度
volatile liquid 挥发性液体
vortex pump 涡流泵
waste liquid 废液
waste water 废水
water absorbent 吸水的
weir 溢流堰

5.4 环境工程

材料阅读

Environment Impact Assessment

1. Introduction

Environment Impact Assessment (EIA) can be defined as the study to predict the effect of a proposed activity/project on the environment. A decision making tool, EIA compares various alternatives for a project and seeks to identify the one which represents the best combination of economic and environmental costs and benefits.

EIA systematically examines both beneficial and adverse consequences of the project and ensures that these effects are taken into account during project design. It helps to identify possible environmental effects of the proposed project, proposes measures to mitigate adverse effects and predicts whether there will be significant adverse environmental effects, even after the mitigation is implemented. By considering environmental effects and mitigation early in the project planning cycle, environmental assessment has many benefits, such as protection of environment, optimum utilisation of resources and saves overall time and cost of the project. Properly conducted EIA also lessens conflicts by promoting community participation, informs decision makers, and helps lay the base for environmentally sound projects. Benefits of integrating EIA have been observed in all stages of a project, from exploration and planning, through construction, operations, decommissioning, and beyond site closure.

2. The EIA Process

The stages of an EIA process will depend upon the requirements of the country or donor. However, most EIA processes have a common structure and the application of the main stages is a basic standard of good practice.

The environment impact assessment consists of eight steps with each step equally important in determining the overall performance of the project. Typically, the EIA process begins with screening to ensure time and resources are directed at the proposals that matter environmentally and ends with some form of follow up on the implementation of the decisions and actions taken as a result of an EIA report. The eight steps of the EIA process is briefly presented below:

• Screening: First stage of EIA, which determines whether the proposed project, requires an EIA and if it requires EIA, then the level of assessment required.

• Scoping: This stage identifies the key issues and impact that should be further investigated. This stage also defines the boundary and time limit of the study.

• Impact analysis: This stage of EIA identifies and predicts likely environmental and social impact of the proposed project and evaluates the significance.

• Mitigation: This step in EIA recommends the actions to reduce and avoid the potential adverse environmental consequences of development activities.

• Reporting: This stage presents the result of EIA in a form of a report to the decision making body and other interested parties.

• Review of EIA: It examines the adequacy and effectiveness of the EIA report and provides information necessary for the decision-making.

• Decision-making: It decides whether the project is rejected, approved or needs further change.

• Post monitoring: This stage comes into play once the project is commissioned. It checks whether the impacts of the project do not exceed the legal standards and implementation of the mitigation measures are in the manner as described in the EIA report.

（摘自 http://www.cseindia.org/programme/industry/eia/introduction_eia.htm#int.）

环境工程常用词汇

absolute pressure 绝对压力
absorption 吸收
absorption field 吸收场
acid 酸
adiabatic lapse rate 绝热递减率
adsorption 吸附作用
advanced treatment 深度处理
acid rain 酸雨
acoustic material 隔音材料
acre-foot 英亩-英尺
activated sludge 活性污泥
aeration 曝气
aerobe 好氧生物
aerobic 好氧的

aerosol 气溶胶
air quality index 空气质量指数
algal bloom 水华
algae 藻类
alkalinity 碱度
alum 明矾
ambient sample 环境空气样品
anaerobic 厌氧微生物
anaerobic 厌氧的
anthropogenic 人为的
aquatic organism 水生生物
aquiclude 滞水层
aquifer 含水层
area method 面积法
artesian aquifer 承压含水层
ash 灰烬
atom 原子
autotrophic organisms 自养型生物
backwash 反冲洗
backwater analysis 回水分析
bacteria 细菌
baghouse filter 袋式过滤器
base 碱
base flow 基本流量
batter boards 龙门板撑
bioaugmentation 生物添加技术
biochemical oxygen demand 生化需氧量
biodegradable 可生物降解的
biological treatment 生物处理
bioreactor 生物反应器
bioremediation 生物修复
biosolids 生物固体废物
bioventing 生物通气法
bottom ash 炉底灰
brackish 半碱水
bubbler 气泡式吸收管
bulking sludge 污泥膨胀
carbonaceous material 含碳物质
carcinogenic 致癌的
catalyst 催化剂
catchment area 汇水面积
centrifugal pump 离心泵
cesspool 化粪池
channel flow time 渠道流行时间
chemical oxygen demand 化学需氧量
chemistry 化学
chlorination 加氯消毒
chlorine residual 余氯
chlorofluorocarbons 氯氟烃
clarifier 沉淀池
coagulation 混凝
cogeneration 废热发电
coliform 大肠杆菌
colloid 胶体
combined sewer 合流制污水管
combustible material 可燃物
cometabolism 共代谢
comminutor 粉碎机
communicable disease 传染病
composite runoff coefficient 综合径流系数
composite sample 混合样品
compost 混合肥料
composting 堆肥
compound 化合物
cone of depression 沉降漏斗
corrosive waste 腐蚀性废弃物
cyclone 吸尘器
Darcy's law 达西定律
day-night sound level 昼夜声级
decibel 分贝
decomposition 分解
denitrification 反硝化
desalination 海水淡化
dewatering 脱水

digestion 消化
discharge 排水
disinfection 消毒
dispersed source 分散污染源
drought 干旱
drought flow 旱季流量
dug well 大口井
dump 垃圾堆
dust 尘埃
dustfall bucket 降尘测定桶
ecology 生态学
ecosystem 生态系统
effluent 出水
effluent standards 出水标准
electrostatic precipitator 静电除尘器
element 元素
emissions sampling 排放物取样
environment 环境
ephemeral stream 季节性河流
epidemic 流行病
epilimnion 变温水层
episode 污染事件
eutrophic lake 富营养湖
eutrophication 富营养化
evaporation 蒸发
exfiltration test 渗出检测
extended aeration 延时曝气
facultative bacteria 兼性菌
fecal coliform 粪大肠菌
fecal strep 粪链球菌
filtration 过滤
floc 絮状物
flocculation 絮凝
food web 食物网
force main 污水压力干管
freeboard 超高
freshwater 淡水
fume 烟雾
gage pressure 相对大气压力
garbage 食物垃圾
geomembrane 土工薄膜
global warming 全球变暖
grab sample 瞬时样品
gradually varied flow 渐变流
gravity flow 重力流
greenhouse effect 温室效应
grit chamber 尘砂池
groundwater 地下水
groundwater table 地下水位
hammer mill 锤式粉碎机
hardness 硬度
hazardous waste 有害物质
heavy metals 重金属
Hertz 赫兹
heterotrophic organism 异氧生物
humus 腐殖质
hydraulics 水力学
hydrograph 水文循环
hydrology 水文学
hydrostatic pressure 净水压力
hypolimnion 湖底滞水层
influent 进水
igneous rock 火成岩
ignitable waste 易燃废弃物
Imhoff cone 英霍夫式锥形管
in situ treatment 就地处理
incineration 焚烧
industrial sewage 工业废水
infectious disease 传染病
infiltration 渗入
inflow 流入
inorganic 无机的
interceptor 截留管
ion 离子

ionization 离子化作用
isokinetic sample 等速采样
jar test 烧杯实验
lagoon 氧化塘
land treatment 土地处理
landfill 填埋
laser 激光器
lateral 污水支管
lift station 提升泵站
low-pressure air test 低压空气试验
major losses 管道能量损失
manhole 检查井
mass burning 混烧
membrane filter method 滤膜法
metabolism 新陈代谢
metamorphic rock 变质岩
meteorology 气象学
methane 甲烷
microbar 微巴
microbe 微生物
mist 雾
molecule 分子
most probable number 最大可能数
mulch 覆盖物
municipal solid waste 城市固体废物
mutagenic 引起突变的
nappe 水舌
natural attenuation 自然衰减
natural succession 自然演替
nitrates 硝酸盐
nitrification 硝化
nonferrous metal 有色金属
normal depth 正常水体
nutrient 营养物
oligotrophic lake 贫营养湖
open channel flow 明渠流
organic compound 有机化合物
overland flow 地面径流
oxidation 氧化
oxidation ditch 氧化沟
oxygen sag curve 氧垂曲线
paper-tape sample 纸带取样器
particulate 颗粒物
Pascal 帕
pathogen 致病菌
percolation 渗透
permeability 渗透率实验
permeability 渗透性
photosynthesis 光合作用
pitch 音调
potable water 饮用水
primary treatment 初级处理
protozoa 原生动物
putrefaction 腐烂
pyrolysis 高温分解
radical 原子团
rainfall curves 降雨曲线
rapidly varied flow 急变流
raw sewage 污水原水
reactive waste 危险废物
recharge area 回灌区
recycling 再循环
reduction 还原反应
reformulated gasoline 改良汽油
refuse 垃圾
remediation 修复
respiration 呼吸作用
river basin 江河流域
regulcanization 再硫化
ridgeline 分水线
rubbish 垃圾
runoff 径流
saltwater intrusion 海水入侵
sanitary landfill 卫生填埋

sanitation 环境卫生
screening 筛分
scrubber 洗涤器
secondary pollutant 二级污染物
secondary standards 二级标准
secondary treatment 二级处理
self-purification 自净
septic 厌氧
settling tank 沉淀池
sewage 污水
sewerage 污水系统
shutoff head 静水头
sludge dewatering 污泥脱水
sludge digestion 污泥消化
slurry 泥浆
smoke 烟雾
softening 软化处理
soil series 土系
solid waste 固体废物
sound pressure level 声压级
sparging 喷注
stage 水位
static level 静水位
storm sewage 雨水管
superfund 超级基金
surface water 地表水
suspended solids 悬浮固体
tailwater analysis 尾水分析
terrestrial 陆生的
time of concentration 汇流时间
tipping floor 卸料间
toxic waste 有毒废物
transpiration 蒸腾
trash 垃圾
trophic level 营养级
troposphere 对流层
turnover 翻转
underground injection 地下注射
vacuum filter 真空过滤器
virus 病毒
volatile 挥发性的
waste minimization 废物最小化
water table 地下水位
watershed 流域
weathering 风化
weir 堰
wetland 湿地
working level 工作强度

5.5 生物化学

材料阅读

Biochemistry

Biochemistry is the study of the chemical processes in living organisms. It deals with the structure and function of cellular components, such as proteins, carbohydrates, lipids, nucleic acids, and other biomolecules.

The term carbohydrate reflects the fact that many of the compounds in this category have the empirical formula "CH_2O—", they are literally "hydrates of carbon." Carbohydrates are the primary source of food energy for most living systems. They include simple sugars such as glucose ($C_6H_{12}O_6$) and sucrose ($C_{12}H_{22}O_{11}$) as well as polymers of these sugars such as starch, glycogen, and cellulose. Carbohydrates are produced from CO_2 and H_2O during photosynthesis and are therefore the end products of the process by which plants capture the energy in sunlight.

Lipids, on the other hand, are defined on the basis of their physical properties. Any molecule in a biological system that is soluble in nonpolar solvents is classified as a lipid. The lipid known as cholesterol, for example, is virtually insoluble in water, but it is soluble in a variety of nonpolar solvents,including the nonpolar region between the inner and outer surfaces of a membrane.

The name nucleic acid was originally given to a class of relatively strong acids that were found in the nuclei of cells. As monomers, nucleic acids such as adenosine triphosphate (ATP) are involved in the process by which cells capture food energy and make it available to fuel the processes that keep cells alive. As polymers, they store and process the information that allows the organism to grow and eventually reproduce.

Although there are a vast number of different biomolecules, many are complex and large molecules (called polymers) that are composed of similar repeating subunits (called monomers). Each class of polymeric biomolecule has a different set of subunit types. For example, a protein is a polymer whose subunits are selected from a set of 20 or more amino acids. Biochemistry studies the chemical properties of important biological molecules, like proteins, in particular the chemistry of enzyme-catalyzed reactions.

The biochemistry of cell metabolism and the endocrine system has been extensively described. Other areas of biochemistry include the genetic code (DNA, RNA), protein synthesis, cell membrane transport, and signal transduction.

1. DNA Structure

The base relationship in the dacble helix structure of DNA:

Adenine(A) → purine
Thymine(T) → pyrimidine (DNA only)
Guanine(G) → purine
Cytosine(C) → pyrimidine
Uracil(U) → pyrimidine (RNA only)

Deoxyribonucleic acid (DNA) is a nucleic acid that contains the genetic instructions used in the development and functioning of all known living organisms and some viruses (Figure5.4). The main role of DNA molecules is the long-term storage of information. DNA is often compared to a set of blueprints or a recipe, or a code, since it contains the instructions needed to construct other components of cells, such as proteins and RNA molecules. The DNA segments that carry this genetic information are called genes, but other DNA sequences have structural purposes, or are involved in regulating the use of this genetic information.

Chemically, DNA consists of two long polymers of simple units called nucleotides, with backbones made of sugars and phosphate groups joined by ester bonds. These two strands run in opposite directions to each other and are therefore anti-parallel. Attached to each sugar is one of four types of molecules called bases. It is the sequence of these four bases along the backbone that encodes information. This information is read using the genetic code, which specifies the sequence of the amino acids within proteins. The code is read by copying stretches of DNA into the related nucleic acid RNA, in a process called transcription.

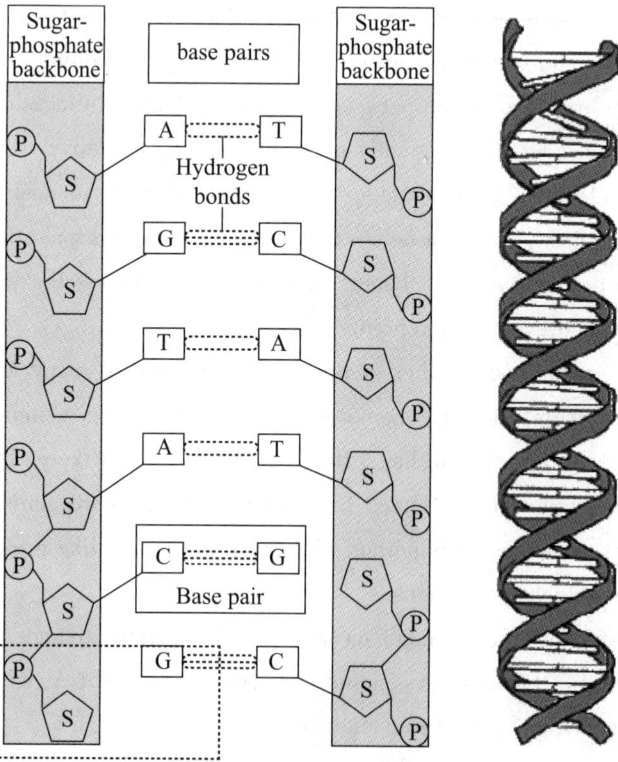

Figure 5.4 DNA structure
图5.4 DNA结构

Within cells, DNA is organized into structures called chromosomes. These chromosomes are duplicated before cells divide, in a process called DNA replication. Eukaryotic organisms (animals, plants, fungi, and protists) store their DNA inside the cell nucleus, while in prokaryotes (bacteria and archae) it is found in the cell's cytoplasm. Within the chromosomes, chromatin proteins such as histones compact and organize DNA. These compact structures guide the interactions between DNA and other proteins, helping control which parts of the DNA are transcribed.

2. Prontein

A protein is a biological polymer comprising numerous amino acids linked recursively through peptide bonds between a carboxyl group and an amino group of adjacent amino acids to form a long chain with the defining side group of each amino acid protruding from it. The sequence of amino acids in a protein is defined by a gene and encoded in the genetic code, which selects protein components from a set of 20 "standard" amino acids.

Protein synthesis is the creation of proteins using DNA and RNA. Proteins can often be synthesized directly from genes by translating mRNA. When a protein is harmful and needs to be available on short notice or in large quantities, a protein precursor is produced. A proprotein is an inactive protein containing one or more inhibitory peptides that can be activated when the inhibitory sequence is removed by proteolysis during posttranslational modification. A preprotein is a form that contains a signal sequence (an N-terminal signal peptide) that specifies its insertion into or through membranes; i.e., targets them for secretion. The signal peptide is cleaved off in the endoplasmic reticulum. Preproproteins have both sequences (inhibitory and signal) still present(Figure 5.5).

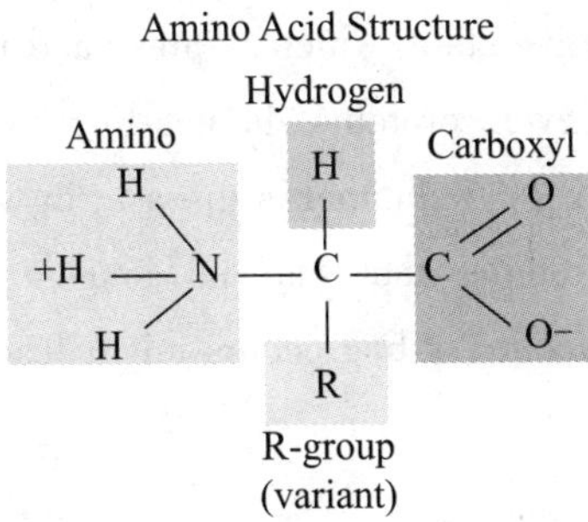

Figure 5.5 Amino acid structure
图5.5 氨基酸结构

For synthesis of protein, a succession of tRNA molecules charged with their appropriate amino acids have to be brought together with an mRNA molecule and matched up by base-pairing through their anticodons with each of its successive codons. The amino acids then have to be linked together to extend the growing protein chain, and the tRNAs, relieved of their burdens, have to be released. This whole complex of processes is carried out by a giant multimolecular machine, the ribosome, formed of two main chains of RNA, called ribosomal

RNA (rRNA), and more than 50 different proteins. This evolutionarily ancient molecular juggernaut latches onto the end of an mRNA molecule and then trundles along it, capturing loaded tRNA molecules and stitching together the amino acids they carry to form a new protein chain.

Transcription-Protein synthesis starts in the nucleus, where the DNA is held. DNA structure is two chains of sugars and phosphates joined by pairs of nucleic acids; Adenine, Guanine, Cytosine, and Thymine. Similar to DNA replication, the DNA is "unzipped" by the enzyme helicase, leaving the single nucleotide chain open to be copied. RNA polymerase reads the DNA strand, and synthesizes a single strand of messenger RNA (mRNA). This single strand of mRNA leaves the nucleus through nuclear pores, and migrates into the cytoplasm where it joins with ribosomes. Where protein synthesis occurs by the formation of peptide bonds and poly peptide chains.

Note: in the new RNA strand, the nucleotide Uracil takes the place of Thymine.

Translation-the process of converting the mRNA codon sequences into an amino acid polypeptide chain.

- Initiation-A ribosome attaches to the mRNA and starts to code at the fMet codon (usually AUG, sometimes GUG or UUG).
- Elongation-tRNA brings the corresponding amino acid to each codon as the ribosome moves down the mRNA strand.
- Termination-Reading of the final mRNA codon (a.k.a. the STOP codon), which ends the synthesis of the peptide chain and releases it.

(1) The four levels of protein structure

Proteins fold into unique three-dimensional structures. The shape into which a protein naturally folds is known as its native state, which is presumed to be determined by its sequence of amino acids. Sometimes, however, proteins do not fold properly. The incorrect folding of proteins can lead to illnesses such as Alzheimer's disease, in which brain function is limited by deposits of incorrectly-folded proteins that can no longer perform their functions. A full understanding of why incorrect protein folding occurs might lead to advances in the treatment of diseases like Alzheimer's.

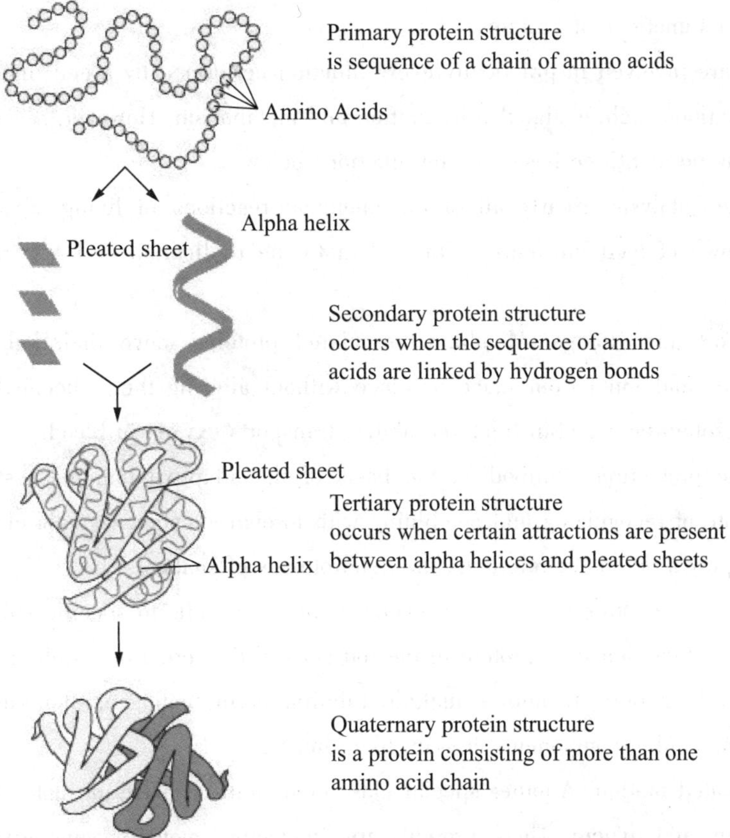
Primary protein structure
is sequence of a chain of amino acids
Amino Acids
Alpha helix
Pleated sheet
Secondary protein structure
occurs when the sequence of amino
acids are linked by hydrogen bonds
Pleated sheet
Tertiary protein structure
occurs when certain attractions are present
between alpha helices and pleated sheets
Alpha helix
Quaternary protein structure
is a protein consisting of more than one
amino acid chain

Figure 5.6　The four levels of protein structure

图5.6　蛋白质结构的4个层次

Biochemists refer to four distinct aspects of a protein's structure(Figure 5.6):

• Primary structure is the linear amino acid sequence encoded by DNA. Any error in this sequence, such as the substitution of one amino acid residue for another, may lead to a congenital disease.

• Secondary structures are highly patterned sub-structures that form in the interaction of amino acid residues near to each other on the chain. The most common are the alpha helix and the beta sheet. There can be many different secondary motifs present in one single protein molecule.

• Tertiary structure refers to the overall, three-dimensional shape of a single protein molecule. This spatial relationship of amino acid residues that are far apart on the sequence is primarily formed by hydrophobic interactions, though hydrogen bonds and ionic interactions, and disulfide bonds are usually involved as well.

• Some proteins may have a quaternary structure, a shape or structure that results from the union of more than one protein molecule (called subunits in this context), which function as part of the larger assembly, or protein complex. Hemoglobin, which serves as an oxygen carrier in blood, has a quaternary structure of four subunits.

(2) Major Functions of Proteins

Proteins are involved in practically every function performed by a cell, including regulation of cellular functions such as signal transduction and metabolism. However, several major classes of proteins may be identified based on the functions below:

• Enzyme catalysis. Nearly all of the chemical reactions in living organisms—from the initial breakdown of food nutrients in the saliva to the replication of DNA—are catalyzed by proteins.

• Transport and storage. Membrane-associated proteins move their substrates (such as small molecules and ions) from place to place without altering their chemical properties. For example, the protein hemoglobin (pictured above) transports oxygen in blood.

• Immune protection. Antibodies, the basis of the adaptive immune system, are soluble proteins capable of recognizing and combining with foreign substances. This class also includes toxins, which play a defensive role (e.g., the dendrotoxins of snakes).

• Signaling. Receptors mediate the responses of nerve cells to specific stimuli. Rhodopsin, for example, is a light sensitive protein in the rod cells of the retina of vertebrates.

• Structural support. Examples include tubulin, actin, collagen, and keratin, which are important strengthening components of skin, hair, and bone.

• Coordinated motion. Another special class of proteins consists of motor proteins such as myosin, kinesin, and dynein. These proteins are "molecular motors," generating physical force which can move organelles, cells, and entire muscles. Proteins are the major components of muscle, and muscle contraction involves the sliding motion of two kinds of protein filaments. At the microscopic level, the propulsion of sperm by flagella is produced by protein assemblies.

• Control of growth and differentiation. In higher organisms, growth factor proteins such as insulin control the growth and differentiation of cells. Transcription factors regulate the activation of transcription in eukaryotes, while cyclins regulate the cell cycle, the series of events in a eukaryotic cell between one cell division and the next.

3. Enzyme

An enzyme is a biological catalyst that regulates the rate of a chemical reaction in a living organism. Most enzymes are proteins, though certain nucleic acids, called ribozymes, are also capable of catalytic activity.

Enzymes are essential to sustain life because most chemical reactions in biological cells, such as the digestion of food, would occur too slowly or would lead to different products without the activity of enzymes. Most inherited human diseases result from a genetic mutation, overproduction, or deficiency of a single critical enzyme. For example, lactose intolerance, the inability to digest significant amounts of lactose, which is the major sugar found in milk, is caused by a shortage of the enzyme lactase.

For an enzyme to be functional, it must fold into a precise three-dimensional shape. How such a complex folding can take place remains a mystery. A small chain of 150 amino acids making up an enzyme has an extraordinary number of possible folding configurations: if it tested 10^{12} different configurations every second, it would take about 10^{26} years to find the right one. Yet, a denatured enzyme can refold within fractions of a second and then precisely react in a chemical reaction. To some, it suggests that quantum effects are at work even at the large distances (by atomic standards) spanned by a protein molecule. At least, it demonstrates a stunning complexity and harmony in the universe.

While all enzymes have a biological role, some enzymes are also used commercially. For instance, many household cleaners use enzymes to speed up the breakdown of protein or starch stains on clothes.

Like all catalysts, enzymes work to lower the activation energy of a reaction, or the initial energy input necessary for most chemical reactions to occur. Heat cannot be added to a living system, so enzymes provide an alternate pathway: they bond with a substrate (the substance involved in the chemical reaction) to form a "transition state," an unstable intermediate complex that requires less energy for the reaction to proceed. Like any catalyst, the enzyme remains unaltered by the completed reaction and can therefore continue to interact with substrates. Enzymes may speed up reactions by a factor of many millions.

Enzymes can be affected by molecules that increase their activity (activators) or decrease their activity (inhibitors). Many drugs act by inhibiting enzymes. Aspirin works by inhibiting CO*x*-1 and CO*x*-2, the enzymes that produce prostaglandin, a hormonal messenger that signals inflammation. By inhibiting the activity of these enzymes, aspirin suppresses our experience of pain and inflammation.

A reaction catalyzed by enzymes must be spontaneous; that is, having a natural tendency to occur without needing an external push. (Thermodynamically speaking, the reaction must contain a net negative Gibbs free energy.) In other words, the reaction would run in the same direction without the enzyme, but would occur at a significantly slower rate. For example, the breakdown of food particles such as carbohydrates into smaller sugar components occurs spontaneously, but the addition of enzymes such as amylases in our saliva makes the reaction occur quickly.

Enzymes can pair two or more reactions, so that a spontaneous reaction can be used to drive an unfavorable one. For example, the cleavage of the high-energy compound ATP is often used to power other, energetically unfavorable chemical reactions, such as the building of proteins.

4. Metabolism

Metabolism could be described as all the chemical and physical reactions that take place in order to maintain growth and normal functioning, as shown in Figure 5.7. It includes anabolism (building up molecules required by the body from nutrients manufactured elsewhere or digested) and catabolism (breaking down molecules in order to obtain energy).

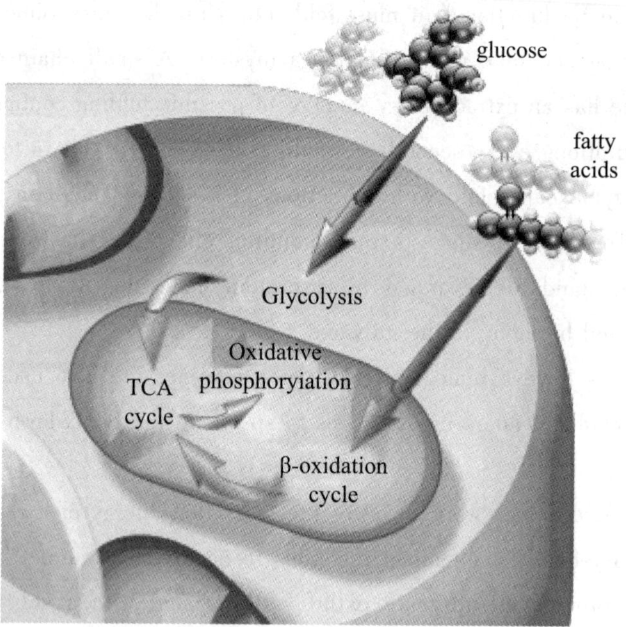

Figure 5.7 Metabolism in the human body
图5.7 人体中的新陈代谢

In terms of catabolism, there are three important processes which the body goes through in order to breakdown glucose and fat from the body. Glucose, obtained from the diet as glucose or in the form of other sugars and carbohydrates, is broken down to pyruvate through the process of glycolysis, and when pyruvate is converted into acetyl CoA, this enters the tricarboxylic acid cycle (TCA cycle).

Fats are found in the diet generally as triacylglycerol which is broken down into its 3 fatty acids. Each of these is converted into a fatty acyl CoA molecule which enters the beta oxidation cycle to produce pyruvate. This can then enter the TCA cycle in the same way as before.

These processes produce chemicals known as NADH (from NAD^+) and $FADH_2$ (from FAD), which enter into a process known as oxidative phosphorylation to produce the all important chemical, adenosine triphosphate. ATP, being so essential, can be derived from various different processes. Those listed above and described in this section are very important, and form the majority of energy-generating reactions, but it's worth remembering that in hunger and starvation, the body employs some excellently designed back-up mechanisms. While essential to normal functioning, the body is not solely limited to this narrow range of activities.

5. Citric Acid Cycle

The citric acid cycle, also known as the Krebs cycle or tricarboxylic acid (TCA) cycle (Figure 5.8), is a series of chemical reactions in the cell that breaks down food molecules into carbon dioxide, water, and energy. In plants and animals, these reactions take place in the mitochondria of the cell as part of cellular respiration. Many bacteria perform the citric acid cycle too, though they do not have mitochondria so the reactions take place in the cytoplasm of bacterial cells. Sir Hans Adolf Krebs, a British biochemist, is credited with discovering the cycle. Sir Krebs outlined the steps of the cycle in 1937.

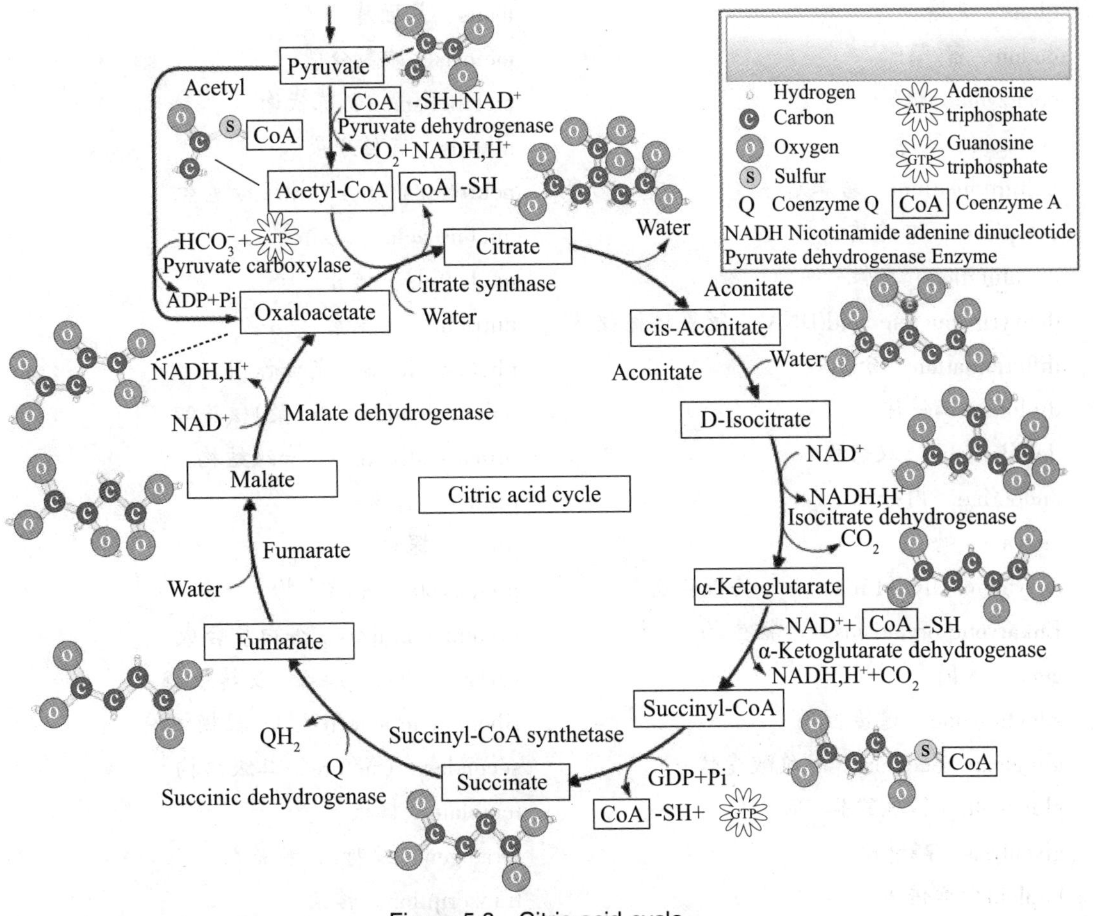

Figure 5.8　Citric acid cycle

图5.8　三羧酸循环

The overall reaction for the citric acid cycle is:

Acetyl-CoA + 3NAD$^+$ + Q + GDP + Pi + 2H_2O $\longrightarrow$ CoA-SH + 3NADH + 3H$^+$ + QH_2 + GTP + 2CO_2

where Q is ubiquinone and Pi is inorganic phosphate.

（摘自　http://en.wikipedia.org/wiki/Main_Page.）

生物化学常用词汇

amino acid　氨基酸

anabolism　合成代谢

antibody　抗体

antigen　抗原

base pair　碱基对

biomolecule　生物分子

catabolism　分解代谢
chromosomes　染色体
clone　克隆
codon　密码子
coenzyme　辅酶
collagen　胶原
culture medium　培养基
cytoplasm　细胞质
denaturation　变性
deoxyribonucleic acid(DNA)　脱氧核糖核酸
differentiation　分化
diploid　二倍体
double helix　双螺旋
endocrine　内分泌
enzyme　酶
enzyme-catalyzed reaction　酶催化反应
Eukaryotic organisms　真核生物
gene　基因
genetic code　遗传密码
glutamate receptors　谷氨酸受体
glutamate　谷氨酸盐
glycolysis　糖酵解
haploid　单倍体
hormone　激素
in vitro　体外的
in vivo　体内的
insulin　胰岛素
locus　基因库
meiosis　减数分裂
metabolism　新陈代谢
mitosis　有丝分裂
neurotransmitter　神经传递素
nucleic acids　核酸
nucleotide　核苷
nutrient　营养素
photosynthesis　光合作用
polynucleotide　多(聚)核苷酸
primary structure　一级结构
primer　引物
probe　探针
prokaryote　原核生物
protein synthesis　蛋白质合成
reverse transcriptase　反转录酶
ribonucleic acid(RNA)　核糖核酸
secondary structure　二级结构
template　模板
teratogen　致畸剂，畸胎剂
transcription　转录
tricarboxylic acid cycle　三羧酸循环

第6章　化学化工专业英语阅读训练

6.1　化学工程导论

阅读 A

What is Chemical Engineering?

In a wider sense, engineering may be defined as a scientific presentation of the techniques and facilities used in a particular industry. For example, mechanical engineering refers to the techniques and facilities employed to make machines. It is predominantly based on mechanical forces which are used to change the appearance and/or physical properties of the materials being worked, while their chemical properties are left unchanged. Chemical engineering encompasses the chemical processing of raw materials, based on chemical and physico-chemical phenomena of high complexity.

Thus, chemical engineering is that branch of engineering which is concerned with the study of the design, manufacture, and operation of plant and machinery in industrial chemical processes.

Chemical engineering is above all based on the chemical sciences, such as physical chemistry, chemical thermodynamics, and chemical kinetics. In doing so, however, it does not simply copy their findings, but adapts them to bulk chemical processing. The principal objectives that set chemical engineering apart from chemistry as a pure science, is "to find the most economical route of operation and to design commercial equipment and accessories that suit it best of all". Therefore, chemical engineering is inconceivable without close ties with economics, physics, mathematics, cybernetics, applied mechanics, and other technical sciences.

In its early days, chemical engineering was largely a descriptive science. Many of the early textbooks and manuals on chemical engineering were encyclopedias of the commercial production processes known at the time. Progress in science and industry has brought with it an impressive increase in the number of chemical manufactures. Today, petroleum for example serves as the source material for the production of about 80 thousand chemicals. The expansion of the chemical process industries on the one hand and advances in the chemical and technical sciences on the other have made it possible to lay theoretical foundations for chemical processing.

As the chemical process industries forged ahead, new data, new relationships and new generalizations were added to the subject-matter of chemical engineering. Many branches in their own right have separated from the main stream of chemical engineering, such as process and plant design, automation, chemical process simulation and modeling, etc.

Historically, chemical engineering is inseparable from the chemical process industries. In its early days chemical engineering which came into being with the advent of early chemical trades was a purely descriptive division of applied chemistry.

The manufacture of basic chemical products on Europe appears to have begun in the 15th century when small, specialized businesses were first set up to turn out acids, alkalis, salts, pharmaceutical preparations, and some organic compounds.

For all the rhetoric of nineteenth-century academic chemists in Britain urging the priority of the study of pure chemistry over applied, their students who became works chemists were little more than qualitative and quantitative analysts. Before the 1880s this was equally true of German chemical firms, who remained content to retain academic consultants who pursued research within the university and who would occasionally provide the material for manufacturing innovation. By the 1880s, however, industrialists were beginning to recognize that the scaling up of consultants' laboratory preparations, and syntheses was a distinctly different activity from laboratory investigation. They began to refer to this scaling problem and its solution as "chemical engineering",possibly because the mechanical engineers who had already been introduced into works to who seemed best able to understand the process involved. The academic dichotomy of head and hand died slowly.

Unit operation. In Britain when in 1881 there was an attempt to name the new society of chemical industry as the "society of chemical engineers", the suggestion was turned down. On the other hand, as a result of growing pressure from the industrial sector the curricula of technical institutions began to reflect, at last, the need for chemical engineers rather than competent analysts. No longer was mere description of existing industrial processes to suffice. Instead the expectation was that the processes generic to various specific industries would be analyzed, thus making room for the introduction of thermodynamic perspectives, as well as those being opened up buy the new physical chemistry of kinetics, solutions and phases.

A key figure in this transformation was the chemical consultant, George Davis (1850~1907), the first secretary of the Society of Chemical Industry. In 1887 Davis, then a lecture at the Manchester Technical School, gave a series of lectures on chemical engineering, which he defined as the study of "the application of machinery and plant to the utilization of chemical action on the large scale". The course, which revolved around the type of plant involved in large-scale industrial operations such as drying, crashing, distillation, fermentation, evaporation and crystallization, slowly became recognized as a model for courses elsewhere, not only in Britain, but overseas. The first fully fledged course in chemical engineering in Britain was not introduced until 1909; though in America, Lewis Norton (1855~1893) of MIT pioneered a Davis-type course as early as 1888.

In 1915, Arthur D. Little, in a report on MIT's program, referred to it as the study of "unit operations" and this neatly encapsulated the distinctive feature of chemical engineering in the twentieth century. The reasons for the success of the Davis movement are clear: it avoided revealing the secrets of specific chemical processes protected by patents or by an owner's reticence—factors that had always seriously inhibited manufacturers from supporting academic programs of training in the past. Davis overcame this difficulty by converting chemical industries "into separate phenomena which could be studied independently" and, indeed, experimented with in pilot plants within a university or technical college workshop.

During this period of intensive development of unit operations, other classical tools of chemical engineering analysis were introduced or were extensively developed. These included studies of the material and energy balance of processes and fundamental thermodynamic studies of multicomponent systems.

Chemical engineers played a key role in helping the United States and its allies win World War II. They developed routes to synthetic rubber to replace the sources of natural rubber that were lost to the Japanese early in the war. They provided the uranium-235 needed to build the atomic bomb, scaling up the manufacturing process in one step from the laboratory to the largest industrial plant that had ever been built. And they were instrumental in perfecting the manufacture of penicillin, which saved the lives of potentially hundreds of thousands of wounded soldiers.

The Engineering Science Movement. Dissatisfied with empirical descriptions of process equipment performance, chemical engineers began to reexamine unit operations from a more fundamental point of view. The phenomena that take place in unit operations were resolved into sets of molecular events. Quantitative mechanistic models for these events were developed and used to analyze existing equipment. Mathematical models of processes and reactors were developed and applied to capital-intensive U. S. industries such as commodity petrochemicals.

Parallel to the growth of the engineering science movement was the evolution of the core chemical engineering curriculum in its present form. Perhaps more than any other development, the core curriculum is responsible for the confidence with which chemical engineers integrate knowledge from many disciplines in the solution of complex problems.

Check Your Understanding

Exercise 1 Mark the following statements with T(true) or F(false) according to the passage.

(1) Mechanical engineering is predominantly based on mechanical forces which are used to change both the physical properties and the chemical properties of the materials. ( )

(2) Chemical engineering has close ties with economics, physics, mathematics, cybernetics, applied mechanics, and other technical sciences. ( )

(3) The development of science and industry has increased the chemical manufactures. ( )

(4) In history, chemical engineering is independent from the chemical industries. ( )

(5) Academic chemists in the 19th century focused on applied chemistry instead of pure chemistry. ( )

(6) According to George Davis, chemical engineering is defined as the study of "the application of machinery and plant to utilization of chemical action on the large scale". ( )

(7) The concept of unit operation was created in 1909. ( )

(8) The chemical engineering curriculum failed to grow with the development of engineering science. ( )

Exercise 2 Give brief answers to the following questions.

(1) What is engineering in broad terms?

(2) What is the foundation of chemical engineering?

(3) What laid theoretical foundations for chemical processing?

(4) Why thermodynamics as well as those being opened up buy the new physical chemistry of kinetics, solutions and phases.

Exercise 3 Complete the following sentences by translating the Chinese given in the brackets.

（1）For example, mechanical engineering refers to（机器制造的技术和设备）。

（2）In doing so, however, it does not simply copy their findings,（而是使他们适应批量化学处理）。

（3）In its early days,（化学工程在很大程度上是一门描述性科学）。

（4）Chemical engineering is above all based on the chemical sciences,（例如物理化学、化学热力学和化学动力学）。

Exercise 4　Translate the following sentences into English.

（1）在更广泛的意义上，工程学可以被定义为对特定行业中使用的技术和设施进行科学展示的学科。

（2）它不只是简单地复制他们的发现，而是使之适应于大量的化学处理。

（3）科学和工业的进步推动了化工产品数量的增加。

（4）今天，石油是大约8万种化学品的生产原料。

Exercise 5 Translate the following passage into Chinese.

During this period of intensive development of unit operations, other classical tools of chemical engineering analysis were introduced or were extensively developed. These included studies of the material and energy balance of processes and fundamental thermodynamic studies of multicomponent systems.

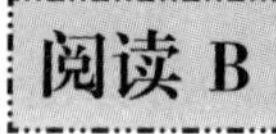

Basic Trends in Chemical Engineering

Over the next few years, a confluence of intellectual advances, technologic challenges, and economic driving forces will shape a new model of what chemical engineering is and what chemical engineering do.

The focus of chemical engineering has always been industrial processes that change the physical state or chemical composition of materials. Chemical engineers engage in the synthesis, design, testing scale-up, operation, control and optimization of these processes. The traditional level of size and complexity at which they have worked on these problems might be termed the mesoscale. Examples of this scale include reactors and equipment for single processes (unit operations) and combinations of unit operations in manufacturing plants. Future research at the mesoscale will be increasingly supplemented by dimensions—the microscale and the dimensions of extremely complex systems—the macroscale.

Chemical engineers of the future will be integrating a wider range of scales than any other branch of engineering. For example, some may work to relate the macroscale of the environment to the mesoscale of combustion systems and the microscale of molecular reactions and transport. Other may work to relate the macroscale performance of a composite aircraft to the mesoscale chemical reactor in which the wing was formed, the design of the reactor perhaps having been influenced by studies of the microscale dynamics of complex liquids.

Thus, future chemical and engineers will conceive and rigorously solve problems on a continuum of scales ranging from microscale. They will bring new tools and insights to research and practice from other disciplines: molecular biology, chemistry, solid-state physics, materials science, and electrical engineering. And they will make increasing use of computers, artificial intelligence, and expert system in problem solving, in product and process design, and in manufacturing.

Two important development will be part of this unfolding picture of the discipline.

Chemical engineers will become more heavily involved in product design as a complement to process design. As the properties of a product in performance become increasingly linked to the way in which it is processed, the traditional distinction between product and process design will become blurred. There will be a special design challenge in established and emerging industries that produce proprietary, differentiated products tailored to exacting performance specifications. These products are characterized by the need for rapid innovatory ad they are quickly superseded in the marketplace by newer products.

Chemical engineers will be frequent participants in multidisciplinary research efforts. Chemical engineering has a long history of fruitful interdisciplinary research with the chemical sciences, particularly industry. The position of chemical engineering as the engineering discipline with the strongest tie to the molecular sciences is an asset, since such sciences as chemistry, molecular biology, biomedicine, and solid-state physics are providing the seeds for tomorrow' s technologies. Chemical engineering has a bright future as the "interfacial discipline", that will bridge science and engineering in the multidisciplinary environments where these new technologies will be brought into being.

Check Your Understanding

Exercise 1　Mark the following statements with T(true) or F(false) according to the passage.

(1) Future chemical and engineers will conceive and rigorously solve problems on a continuum of the mesoscale.(　)

(2) The products are characterized by the need for rapid innovatory AD.(　)

(3) The traditional distinction between product and process design will become clear.(　)

(4) Two important development will be part of this unfolding picture of the discipline.(　)

Exercise 2　Give brief answers to the following questions.

(1) What is the focus of chemical engineering?

(2) What is termed the mesoscale?

(3) What does chemical engineering do?

(4) Why do products are characterized by the need for rapid innovatory ad?

Exercise 3 Complete the following sentences by translating the Chinese given in the brackets.

(1) Chemical engineers of the future(将形成比其他任何工程部门更大的规模。)

(2) The design of the reactor perhaps having been influenced(通过研究复杂液体的微尺度动力学)。

(3) Chemical engineers will become more heavily involved in product design(作为工艺设计的补充)。

(4) The position of(化学工程作为与分子科学关系最紧密的工程学科) to the molecular sciences is an asset.

Exercise 4 Translate the following sentences into English.

(1) 知识的进步、技术挑战和经济驱动力的汇合将形成一个新的模式。

(2) 产品在性能方面的属性与加工方式的联系越来越密切。

(3) 反应堆的设计可能受到复杂液体微尺度动力学研究的影响。

(4) 化学工程将把科学与工程在多学科环境中架起桥梁。

阅读 C

What is Chemistry?

Chemistry is a basic science whose central concerns are :

- the structure and behaviour of atoms (elements).
- the composition and properties of compounds.
- the reactions between substances with their accompanying.energy exchange.
- the laws that unite these phenomena into a comprehensivesystem.

Chemistry is not an isolated discipline, for it merges into physics and biology.The origin of the term is obscure. Chemistry evolved from the medieval practice of alchemy.Its bases were laid by such men as Boyle, Lavoisier, Berzelius, Dalton and Pasteur.

What is Chemical Engineering?

Chemical Engineering is a discipline influencing numerous areas of technology.In broad terms, chemical engineers are responsible for the conception and design of processes for the purpose of production, transformation and transport of materials. This activity begins with experimentation in the laboratory and is followed by implementation of the technology to full-scale production.

The large number of industries which depend on the synthesis and processing of chemicals and materials place the chemical engineer in great demand. In addition to traditional examples such as thechemical, energy and oil industries, opportunities in biotechnology, pharmaceuticals, electronic devicefabrication, and environmental engineering are increasing. The unique training of the chemical engineer becomes essential in these areas whenever processes involve the chemical or physical transformation of matter. For example, chemical engineers working in the chemical industry investigate the creation of new polymeric materials with important electrical, optical or mechanical properties. This requires attention not only to the synthesis of the polymer, but also to the flow and forming processes necessary to create a final product. In biotechnology, chemical engineers have responsibilities in the design of production facilities to use microorganisms and enzymes to synthesize new drugs. Problems in environmental engineering that engage chemical engineers include the development of processes(catalytic converters, effluent treatment facilities) to minimize the release of or deactivate products harmful to the environment.

To carry out these activities, the chemical engineer requires a complete and quantitative under-standing of both the engineering and scientific principles underlying these technological processes. This is reflected in the curriculum of the chemical engineering department which includes the study of applied mathematics, material and energy balances, thermodynamics, fluid mechanics, energy and mass transfer, separations technologies, chemical reaction kinetics and reactor design, and process design. These courses are built on a foundation in the sciences of chemistry, physics and biology.

Chemical Engineering at Stanford

The Department of Chemical Engineering at Stanford is young. in fact, it is barely 40 years old. Within a decade of its inception, however, the department achieved national and international standing in terms of both teaching and research, and is currently ranked among the finest chemical engineering graduate programs in this country. A broad range of research interests is represented in the depart-ment, including both traditional and emerging areas of chemical engineering. The excellence of our graduate program is reflected in the success of our graduates in both industry and academia. Several aspects of the Stanford Chemical Engineering program make it a special place to pursue a Ph.D.Unlike most of the other top programs in the United States, our department is relatively small with twelve faculty and about seventy Ph.D. students and thirty masters students. Partly because of our small size, we have been able to develop a strong sense of community within the department. The small student-to-faculty ratio and the mechanisms that we have established to support and monitor student progress assure that each graduate student receives strong mentoring throughout their timehere. Our emphasis is on graduate training at the Ph. D. level, characterized by modern, forward-look-ing, and challenging research projects. Our research efforts are fundamental in nature, that is, each of our faculty focuses on understanding the basic chemical, physical, and biological phenomena that undertlie the engineering research problem under consideration. Many of our graduate students work on research projects that involve collaboration with researchers from other fields who come from other departments at Stanford or from industry.

Check Your Understanding

Exercise 1　Mark the following statements with T (true) or F (false) according to the passage.

(1) Chemistry, a basic science, is not isolated from other disciplines like physics and biology.(　)

(2) Chemistry mainly concems the structure and properties of atoms and compounds.(　)

(3) Chemical Engineering is influencing numerous areas of technology.(　)

(4) The term chemistry obviously originated from alchemy.(　)

(5) Modem chemical engineers need unique training in modem industries.(　)

(6) Chemicalengineers studying biotechnology have responsibilities in synthesizing new drugs.(　)

(7) The Department of Chemical Engineering at Stanford has a history of less than 40 years.(　)

(8) The research projects at the Ph.D. level are modem, fonward-looking, and challenging. (　)

Exercise 2　Give brief answers to the following questions.

(1) What does chemistry focus on according to the passage?

(2) Generally speaking, what do chemical engineers do?

(3) Why are the chemical engineers greatly demanded?

(4) In what areas is the unique training of the chemical engineer essential?

(5) What can chemical engineers do in environmental engineering?

Exercise 3 Translate the following phrases into English or Chinese.

English	Chinese
______________________	原子的行为特征
the composition of compounds	______________________
the reactions between substances	______________________
______________________	将这些现象纳入到统一体系中并
______________________	融入物理和生物
to carry out these activities	______________________
______________________	化学品的合成与加工处理
the chemical transformation of matter	______________________

Exercise 4 Complete the following sentences by translating the Chinese given in the brackets.

(1) As a basic science, ________________________________(化学主要关注的是四个方面)。(concern)

(2) Chemistry is a discipline ________________________________(融入到物理学和生物学之中)。(merge into)

(3) The bases of chemistry ________________________________(是由下列著名化学家打下的)Boyle, Lavoisier, Berzelius, Dalton and Pasteur. (lay)

(4) Chemical Engineering is ________________________________(一门影响众多技术领域的学科)。(discipline)

(5) The large number of chemical industries ________________________________(导致对化学工程师的需求大增)。(place ... in great demand)

(6) ________________________________(这就要求不仅要注意聚合物的合成而且要注意)the flow and foming processes necessary to create a tinal product. (synthesis)

Exercise 5 Translate the following sentences into English.

(1) 化学是一门基础科学,但是它不是一门孤立的学科,因为它融入了物理学和生物学。

(2) 通常化学工程师要处理环境工程方面的问题。

(3) 斯坦福大学化学工程系目前排名处于美国最佳化学工程研究生点之列。

(4) 我们有许多研究生与斯坦福大学其他专业的或来自工业界的研究人员合作研究项目。

6.2 化工基础知识

阅读 A

Elements and Compounds

Elements are pure substances that can not be decomposed into simpler substances by ordinary chemical changes. At the present time there are 109 known elements. Some common elements that are familiar to you are carbon, oxygen, aluminum, iron, copper, nitrogen and gold. The elements are the building blocks of matter just as the numerals 0 through 9 are the building blocks for numbers. To the best of our knowledge, the elements that have been found on the earth also comprise the entire universe.

About 85% of the elements can be found in nature, usually combined with other elements in minerals and vegetable matter or in substances like water and carbon dioxide. Copper, silver, gold, and about 20 other elements can be found in highly pure forms. Sixteen elements are not found in nature; they have been produced in generally small amounts in nuclear explosions and nuclear research. They are man-made elements.

Pure substances composed of two or more elements are called compounds. Because they contain two or more elements, compounds, unlike elements, are capable of being decomposed into simpler substances by chemical changes. The ultimate chemical decomposition of compounds produces the elements from which they are made.

The atoms of the elements in a compound are combined in whole number ratio, not in fractional parts of an atom. Atoms combined with one another to form compounds which exist as either molecule or ions. A molecule is a small, uncharged individual unit of a compound formed by the union of two or more atoms, if we subdivide a drop of water into smaller and smaller particals, we ultimately obtain a single unit of water known as a molecule of water. This water molecule consists of two hydrogen atoms and one oxygen atom bonded together. We cannot subdivide this unit further without destroying the molecule, breaking it up into its elements. Thus, a water molecule is the smallest unit of the compound water.

An ion is a positive or negative electrically charged atom or group of atoms. The ions in a compound are held together in a crystalline structure by the attractive forces of their positive and negative charges. Compounds consisting of ions do not exist as molecules. Sodium chloride is an example of a non-molecular compound. Although this type of compound consists of large numbers of positive and negative ions, its formula is usually represented by the simplest ratio of the atoms in the compound. Thus, the ratio of ions in sodium chloride is one sodium ion to one chlorine ion.

Compounds exist either as molecules which consist of two or more elements bonded together or in the form of positive and negative ions held together by the attractive force of their positive and negative charges.

The compound carbon monoxide (CO) is composed of carbon and oxygen in the ratio of one atom of carbon to one atom of oxygen. Hydrogen chloride (HCl) contains a ratio of one atom of hydrogen to one atom of chlorine. Compounds may contain more than one atom of the same element. Methane ("natural gas" CH_4) is composed of carbon and hydrogen in a ratio of one carbon atom to four hydrogen atoms; ordinary table sugar (sucrose, $C_{12}H_{22}O_{11}$) contains a ratio of 12 atoms of carbon to 22 atoms of hydrogen to 11 atoms of oxygen. These atoms are held together in the compound by chemical bonds.

There are over three million known compounds, with no end in sight as to the number that can and will be prepared in the future. Each compound is unique and has characteristic physical and chemical properties. Let us consider in some detail two compounds-water and mercuric oxide. Water is a colorless, odorless, tasteless liquid that can be changed to a solid, ice, at 0 ℃ and to a gas, steam at 100 ℃. It is composed of two atoms of hydrogen and one atom of oxygen per molecule, which represents 11.2 percent hydrogen and 88.8 percent oxygen by weight. Water reacts chemically with sodium to produce hydrogen gas and sodium hydroxide, with lime to produce calcium hydroxide, and with sulfur trioxide to produce sulfuric acid. No other compound has all these exact physical and chemical properties, they are characteristic of water alone.

Mercuric oxide is a dense, orange-red powder composed of a ratio of one atom of mercury to one atom of oxygen. Its composition by weight is 92.6 percent mercury and 7.4 percent oxygen, when it is heated to temperatures greater than 3600 ℃, a colorless gas, oxygen, and a silvery liquid metal, mercury, are produced. Here again are specific physical and chemical properties belonging to mercuric oxide and to no other substance. Thus, a compound may be identified and distinguished from all other compounds by its characteristic properties.

Check Your Understanding

Exercise 1　Mark the following statements with T(true) or F(false) according to the passage.

(1) Elements can be decomposed into simpler substances by ordinary chemical changes.(　)

(2) About 85% of the elements are usually combined with other elements in minerals and vegetable matter or in substances.(　)

(3) 20 elements are produced in generally small amounts in nuclear explosions and nuclear research.(　)

(4) A water molecule consists of two hydrogen atoms and one oxygen atom.(　)

(5) The ratio of ions in sodium chloride is one sodium ion to one chlorine ion.(　)

(6) Compounds only contain one atom of the same element.(　)

Exercise 2　Give brief answers to the following questions.

(1) What is the important use of elements?

(2) What is the main difference between elements and compounds?

(3) Why the water molecule is the smallest unit of the compound water?

(4) In what forms do the compounds exist?

(5) What can be produced if water reacts chemically with lime?

Exercise 3 Complete the following sentences by translating the Chinese given in the brackets.

(1) Elements are pure substances that(不能被普通化学反应分解成简单物质的纯净物)。

(2) Compounds are capable of (能被化学反应分解成单质)。

(3) A molecule is a small, uncharged individual unit of (两个或多个原子组成的)。

(4) Although this type of compound consists of large numbers of positive and negative ions, its formula is(通常以在化合物中原子的最简单的比率为代表)。

(5) The compound carbon monoxide (是由碳和氧按1:1的比例组成的)。

Exercise 4 Translate the following sentences into English.

(1) 如果我们把水分成非常微小的颗粒,我们最后就会得到被称为水分子的水的单一单元。

(2) 氯化氢由一个氢原子和一个氯原子组成。

(3) 水是无色、无味的液体，它在0 ℃变成冰，100 ℃变成水蒸气。

(4) 已经有超过300万种已知的化合物，未来将要被制造出来的化合物数目更是数不胜数。

(5) 氧化汞的质量组成是92.6%的汞和7.4%的氧，当把它加热到超过3600 ℃时，会产生无色的气体氧和液体汞。

阅读 B

Types of Chemical Reactions

1. Concept

Chemical reaction is a process that leads to the transformation of one set of chemical substances to another. Chemical reactions can be either spontaneous, requiring no input of energy, or non-spontaneous, often coming about only after the input of some type of energy, heat, light or electricity. Classically, chemical reactions encompass changes that strictly involve the motion of electrons in the forming and breaking of chemical bonds, although the general concept of a chemical reaction, in particular the notion of a chemical equation, is applicable to transformations of elementary particles, as well as nuclear reactions.

The substance/substances initially involved in a chemical reaction are called reactants. Chemical reactions are usually characterized by a chemical change, and they yield one or more products, which usually have properties different from the reactants.

Different chemical reactions are used in combination in chemical synthesis in order to get a desired product. In biochemistry, series of chemical reactions catalyzed by enzymes form metabolic pathways, by which syntheses and decompositions ordinarily impossible in conditions within a cell are performed.

Chemical reactions involve the rupture or rearrangement of the bonds holding atoms together, never atomic nuclei. The total mass and number of atoms of all reactants equals those of all products, and energy is almost always consumed or liberated. Understanding their mechanisms lets chemists alter reaction conditions to optimize the rate or the amount of a given product; the reversibility of the reaction and the presence of competing reactions and intermediate products complicate these studies. Reactions can be syntheses, decompositions, or rearrangements, or they can be additions, eliminations, or substitutions. Examples include oxidation-reduction, polymerization, ionization, combustion(burning), hydrolysis, and acid-base reactions.

Chemical change requires a chemical reaction, a process whereby the chemical properties of a substance are altered by a rearrangement of the atoms in the substance. Of course we cannot see atoms with the naked eye, but fortunately, there are a number of clues that tell us when a chemical reaction has occurred. In many chemical reactions, for instance, the substance may experience a change of state or phase—as for instance when liquid water turns into gaseous oxygen and hydrogen as a result of electrolysis.

Essentially, a chemical reaction is the result of collisions between molecules. According to this collision model, if the collision is strong enough, it can break the chemical bonds in the reactants, resulting in a rearrangement of the atoms to form products. The more the molecules collide, the faster the reaction. Increase in the numbers of collisions can be produced in two ways: either the concentrations of the reactants are increased, or the temperature is increased. In either case, more molecules are colliding.

Increases of concentration and temperature can be applied together to produce an even faster reaction, but rates of reaction can also be increased by use of a catalyst, a substance that speeds up the reaction without participating in it either as a reactant or product. Catalysts are thus not consumed in the reaction. One very important example of a catalyst is an enzyme, which speeds un complex reactions in the human body. At ordinary body temperatures, these reactions are too slow, but the enzyme hastens them along. Thus human life can be said to depend on chemical reactions aided by a wondrous form of catalyst.

2. Reaction Types

Some common kinds of reactions are listed below. Note that it is perfectly possible for a single reaction to fall under more than one category:

(1) Isomerisation, in which a chemical compound undergoes a structural rearrangement without any change in its net atomic composition;Direct combination or synthesis, in which 2 or more chemical elements or compounds unite to form a more complex product:

$$N_2 + 3H_2 \longrightarrow 2NH_3$$

(2) Chemical decomposition or analysis, in which a compound is decomposed into smaller compounds or elements:

$$2H_2O \longrightarrow 2H_2 + O_2$$

(3) Single displacement or substitution, characterized by an element being displaced out

of a compound by a more reactive element:

$$2Na(s)+2HCl(aq) \longrightarrow 2NaCl(aq)+H_2(g)$$

(4) Metathesis or double displacement reaction, in which two compounds exchange ions or bonds to form different compounds:

$$NaCl(aq)+AgNO_3(aq) \longrightarrow NaNO_3(aq)+AgCl(s)$$

Acid-base reactions, broadly characterized as reactions between an acid and a base, can have different definitions depending on the acid-base concept employed. Some of the most common are:

Arrhenius definition: Acids dissociate in water releasing H_3O^+ ions; bases dissociate in water releasing OH^- ions.

Brønsted-Lowry definition: Acids are proton (H^+) donors; bases are proton acceptors, Including the Arrhenius definition.

Lewis definition: Acids are electron-pair acceptors; bases are electron-pair donors. Includes the Brønsted-Lowry definition.

Redox reactions, in which changes in oxidation numbers of atoms in involved species occurred; those reactions can often be interpreted as transference of electrons between different molecular sites or species. An example of a redox reaction is:

$$2S_2O_3^{2-}(aq)+I_2(aq) \longrightarrow S_4O_6^{2-}(aq)+2I^-(aq)$$

In which I_2 is reduced to I^- and $S_2O_3^{2-}$ (thiosulfate anion) is oxidized to $S_4O_6^{2-}$.

Combustion, a kind of redox reaction in which any combustible substance combines with an oxidizing element, usually oxygen, to generate heat and form oxidized products. The term combustion is usually used for only large-scale oxidation of whole molecules, i.e. a controlled oxidation of a single functional group is not combustion.

$$C_{10}H_8+12O_2 \longrightarrow 10CO_2+4H_2O$$

$$CH_2S+6F_2 \longrightarrow CF_4+2HF+SF_6$$

Disproportionation is a redox reaction in which one reactant forming two distinct products varying in oxidation state.

$$2Sn^{2+} \longrightarrow Sn+Sn^{4+}$$

Organic reactions encompass a wide assortment of reactions involving compounds which have carbon as the main element in their molecular structure. The reactions in which an organic compound may take part are largely defined by its functional groups.

Check Your Understanding

Exercise 1　Translate the following phrases into English or Chinese.

催化剂	Spontaneous
可燃物质	Proton acceptors
还原剂	Structural rearrangement
原子氧化数	Reversibility

Exercise 2 Give brief answers to the following questions.

(1) How does the non-spontaneous chemical reaction come about?

(2) In biochemistry, how are the synthesis and decompositions are performed?

(3) Can you list some clues that demonstrate a chemical reaction has occurred?

(4) What is the result if the collisions between molecule are strong enough?

(5) What is a catalyst?

(6) What is the use of an enzyme?

Exercise 3 Fill in the blanks with the correct words.

(1) Chemical reactions can be either____or____.

(2) The general concept of a chemical reaction is applicable to transformations of ____ and _____.

(3) The substances initially involved in a chemical reaction are called _____.

(4) Chemical reactions involve the ____ or _____ of the bonds holding atoms together.

(5) Either the ______ of the reactants increased or the ______is increased will increase the numbers of collisions.

Exercise 4 Translate the following sentences into English.

(1) 通过了解它们的机理,改变反应条件。

(2) 提高浓度和温度都可以加快反应。

(3) 化学反应可以产生单一或多种性质不同于反应物的产物。

(4) 一种有机化合物是否能参加反应很大程度上取决于它的官能团。

阅读 C

Inorganic and Organic Compounds

1. Inorganic Compounds

Traditionally, inorganic compounds are considered to be of a mineral, not biological origin. Complementarily, most organic compounds are traditionally viewed as being of biological origin. Over the past century, the precise classification of inorganic vs organic compounds has become less important to scientists, primarily because the majority of known compounds are synthetic and not of natural origin. Furthermore, most compounds considered the purview of modern inorganic chemistry contain organic ligands.

Many compounds that contain carbon, are considered inorganic; for example, carbon monoxide, carbon dioxide, carbonates, cyanides, cyanates, carbides. In general, however, the workers in these areas are not concerned about strict definitions. Examples range from substances that are strictly inorganic, such as $[Co(NH_3)]Cl_3$, to organometallic compounds such as $Fe(C_5H_5)_2$ and extending to bioinorganic compounds, such as the hydrogenase enzymes.

Minerals are mainly oxides and sulfides, which are strictly inorganic. In fact, most of the earth and the universe is inorganic. Although the components of the Earth's crust are well elucidated, the processes of mineralization and the composition of the deep mantle remain active areas of investigation. Any substance in which two or more chemical elements other than carbon are combined, nearly always in definite proportions, as well as some compounds containing carbon but lacking carbon-carbon bonds (e.g., carbonates, cyanides).

Inorganic compounds may be classified by the elements or groups they contain (e. g., oxides, sulfates). The major classes of inorganic polymers are silicones, silanes, silicates, and borates. Coordination compounds (or complexes), an important subclass of inorganic compounds, consist of molecules with a central metal atom (usually a transition element) bonded to one or more nonmetallic ligands (inorganic, organic, or both).

2. Organic Compounds

Substance whose molecules contain one or more (often many more) carbon atoms (excluding carbonates, cyanides, carbides,) are called organic compound. Until 1828, scientists believed that organic compounds could be formed only by life processes (hence the name). Since carbon has a far greater tendency to form molecular chains and rings than do other elements, its compounds are vastly more numerous (many millions have been described) than all others known. Living organisms consist mostly of water and organic compounds: proteins, carbohydrates, fats, nucleic acids, hormones, vitamins, and a host of others. Natural and synthetic fibres and most fuels, drugs, and plastics are organic. Hydrocarbons contain only carbon and hydrogen; organic compounds with other functional groups include carboxylic acids, alcohols, aldehydes, ketones, phenols, ethers, esters, and other more complexmolecules, including heterocyclic compounds, isoprenoids, and amino acids.

Organic compounds can be found in nature or they can be synthesized in the laboratory. An organic substance is not the same as a "natural" substance. A natural material means that it is essentially the same as it was found in nature, but "organic" means that it is carbon based. A printed circuit board is an example of an organic substrate because the laminate material is made of glass fibers in an epoxy, and epoxies are carbon based.

An organic compound is any member of a large class of chemical compounds whose molecules contain carbon. There is no "official" definition of an organic compound. Some text books define an organic compound as one containing one or more C—H bonds; others include C—C bonds in the definition. Others state that if a molecule contains carbon it is organic.

Even the broader definition of "carbon-containing molecules" requires the exclusion of carbon-containing alloys (including steel), a relatively small number of carbon-containing compounds such as metal carbonates and carbonyls, simple oxides of carbon and cyanides, as well as the allotropes of carbon and simple carbon halides and sulfides, which are usually considered to be inorganic.

To summarize: Most carbon-containing compounds are organic, and most compounds with a C—H bond are organic. Not all organic compounds necessarily contain C—H bonds.

Organic compounds may be classified in a variety of ways. One major distinction is between natural and synthetic compounds. Organic compounds can also be classified or subdivided by the presence of heteroatoms, e.g. organometallic compounds which feature bonds between carbon and a metal, and organophosphorus compounds which feature bonds between carbon and a phosphorus.

Another distinction, based upon the size of organic compounds, distinguishes between small molecules and polymers.

Natural compounds refer to those that are produced by plants or animals. Many of these are still extracted from natural sources because they would be far too expensive to be produced artificially. Examples include most sugars, some alkaloids and terpenoids, certain nutrients such as vitamin B12, and in general, those natural products with large or stereo isometrically complicated molecules present in reasonable concentrations in living organisms. Further compounds of prime importance in biochemistry are antigens, carbohydrates, enzymes, hormones, lipids and fatty acids, nucleic acids, proteins, peptides and amino acids, vitamins.

Check Your Understanding

Exercise 1 Translate the following phrases into English or Chinese.

同分异构	organometallic
生物起源	composition
氨基酸	bioinorganic
合成物质	organometallic

Exercise 2 Mark the following statements with T(true) or F(false) according to the passage.

(1) The majority of known compounds are synthetic and not of natural origin. (　)

(2) Most of the earth and the universe is organic. (　)

(3) Substance whose molecules contain one or more carbon atoms are called organic compound. (　)

(4) Organic compounds can be found in nature or they can be synthesized in the laboratory. (　)

Exercise 3 Fill in the blanks with the correct words.

(1) The majority of known compounds are ______ and not of ______.

(2) Minerals are mainly ______ and _______.

(3) Hydrocarbons contain only ________and ______.

(4)__________ refer to those that are produced by plants or animals.

Exercise 4 Translate the following sentences into English.

(1) 有机化合物可以以各种方式分类。

(2) 将碳以外的两种或两种化学元素组合在一起的任何物质,几乎总是以一定的比例结合。

(3) 天然物质意味着它本质上与在自然界中发现的物质相同。

(4) 有机化合物可以在自然界中发现,也可以在实验室中合成。

Exercise 5 Translate the following passage into Chinese.

Hydrocarbons contain only carbon and hydrogen; organic compounds with other functional groups include carboxylic acids, alcohols, aldehydes, ketones, phenols, ethers, esters, and other more complex molecules, including heterocyclic compounds, isoprenoids, and amino acids.

Inorganic compounds may be classified by the elements or groups they contain (e. g., oxides, sulfates). The major classes of inorganic polymers are silicones, silanes, silicates, and borates. Coordination compounds (or complexes), an important subclass of inorganic compounds, consist of molecules with a central metal atom (usually a transition element) bonded to one or more nonmetallic ligands (inorganic, organic or both).

6.3　化工单元操作

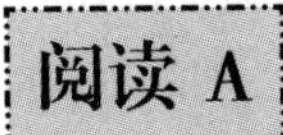

Glossary of the Chemical Process Term

Absorption: A process in which a gas mixture contacts a liquid solvent and a component (or several components) of the gas dissolves in the liquid. In an absorption column or absorption tower, the solvent enters the top of a column, flows down, and emerges at the bottom, and the gas enters at the bottom, flows up (contacting the liquid), and leaves at the top.

Adsorption: A process in which a gas or liquid mixture contacts a solid and a mixture component adheres to the surface of the solid.

Boiler: A process unit in which tubes pass through a combustion furnace, Boiler feed water is fed into the tubes,and heat transferred from the hot combustion products through the tube walls converts the feed water to steam.

Boiling point (at a given pressure): For a pure species, the temperature at which the liquid and vapor can coexist in equilibrium at the given pressure when applied to the heating of a mixture of liquids exposed of liquids exposed to a gas at the given pressure, the temperature at which the mixture begins to boil.

Bottoms product: The product that leaves the bottom of a distillation column. The bottoms product is relatively rich in the less volatile components of the feed to the column.

Bubble point: The temperature at which the first vapor bubble appears when the mixture is heated.

Condensation: A process in which an entering gas is cooled and/or compressed, causing one or more of the gas components to liquefy. Uncondensed gases and liquid condensate leave the condenser as separate streams.

Crystallization: A process in which a liquid solution is cooled, or solvent is evaporated, to an extent that solid crystals of solute form. The crystals in the slurry (suspension of solids in a liquid) leaving the crystallizer may subsequently be separated from the liquid in a filter or centrifuge.

Dew point: The temperature at which the first liquid droplet appears when the mixture is cooled at constant pressure.

Distillation: A process in which a mixture of two or more species is fed to a vertical column that contains either a series of vertically spaced horizontal plates, or solid packing through which fluid can flow. Liquid mixtures of the feed components flow down the column and vapor mixtures flow up. Interphase contact, partial condensation of the vapor, and partial vaporization of the liquid take place throughout the column. The vapor flowing up the column becomes progressively richer in the more volatile components of the feed, and the liquid flowing down becomes richer in the less volatile components. The vapor leaving the top of the column is condensed; part of the condensate is taken off as the overhead product and the rest is recycled to the reactor as reflue becoming the liquid stream that flows down the column. The liquid leaving the bottom of the column is partially vaporized; the vapor is recycled to the reactor as boil up, becoming the vapor stream that flows up the column, and the residual liquid is taken off as the bottom product.

Drying: A process in which a wet solid is heated or contacted with a hot gas stream, causing some or all of the liquid wetting the solid to evaporate. The vapor and the gas evaporate to emerge as one outlet stream, and the solid and remaining residual liquid emerge as a second outlet stream.

Evaporation (vaporization): A process in which a pure liquid, liquid mixture, or solvent in a solution is vaporized.

Extraction (liquid extraction): A process in which a liquid mixture of two species (the solute and the feed carrier) is contacted in a mixer with a third liquid (the solvent) that is immiscible or nearly immiscible with the feed carrier. When the liquids are contacted, solute transfers from the feed carrier to the solvent. The combined mixture is then allowed to settle into two phases that are then separated by gravity in a decanter.

Filtration: A process in which as slurry of solid particles suspended in a liquid passes throu a porous medium. Most of the liquid passes through the medium (e.g., a filter) to form the filtrate, and the solids and some entrained liquid are retained to the filter to form cake. Filtration may also be used to separate solids or liquids from gases.

Heat Energy transferred between a system and its surroundings as a consequence of a temperature difference. Heat always flows from a higher temperature to a lower one.

Heat exchanger: A process unit through which two fluid streams at different temperatures flow opposite sides of a metal barrier. Heat is transferred from the stream at the higher temperature through the barrier to the other stream.

Overhead product: The product that leaves the top of a distillation column. The overhead product is relatively rich in the most volatile components of the feed to the column.

Pump: A device used to propel a liquid or slurry from one location to another, usually through a pipe or tube.

Check Your Understanding

Exercise 1 Put the following into Chinese.

tube wall
given pressure
liquid condensate
residual liquid
interphase contact
transfer from…to…
heat exchanger
liquid crystal

Exercise 2　Put the following into English.

泡点
露点温度
回流
易挥发组织
多空介质
过滤器
塔顶产品
蒸馏柱
结晶器
热交换器

Exercise 3　Give brief answers to the following questions.

(1) What are the definitions point and dew point?

(2) Try to find out some examples of absorption.

(3) Briefly describe the differences between drying, evaporation, extraction and filtration.

(4) The cooling of a crystalline liquid solution or the evaporation of a solvent.

Exercise 4　Mark the following statements with T(true) or F(false) according to the passage.

(1) The temperature at which the mixture begins to boil when applied to a liquid exposed to a gas at a given pressure. (　)

(2) The temperature at which the vapor bubbles appear when the mixture is heated is the bubble point. (　)

(3) The temperature at which the first droplet appears as the mixture cools under constant

pressure is the dew point . (   )

阅读 B

Evaporator and Crystallizer

Alaqua custom designs and manufactures to ASME codes and 3 sanitary standards, a variety of lines of single and multiple effect evaporators: Falling film, forced circulation, long tube vertical, thermally accelerated falling film, mechanical vapor recompression, thermal vapor recompression, wiped film, thin film, submerged combustion, and others. You can run Alaqua's evaporators on steam with high thermal economy, on hot oil or on compressors.

Alaqua's evaporators will improve your overall performance whether;

You are processing food products like milk, proteins, cheese whey, coffee, fruit juices and more.

You are concentrating and crystallizing organic and inorganic chemicals such as: ammonium compounds, sulfates, carbonates, urea, calcium compounds, chlorides, acids, nitrates, pulp and paper, liquors, etc.

You are processing wastewater effluents, waste solvent streams and waste oil products.

Evaporator and crystallizer applications in Chinese and English(partial):

alcohols 乙醇
animal blood 动物血
apple juice 苹果汁
american society of mechanical engineers (ASME) 美国机械工程师协会
boric acid 硼酸
brewer's malt 啤酒麦芽
calcium salts 钙盐
caprolactam 己内酰胺
caustic soda 氢氧化钠
citrus juice coffee 柑橘类果汁咖啡
collagen 胶原
copper sulfate distiller 硫酸铜蒸馏器
distiller slop 蒸馏废油
extract 提取物
ethanol stillage 乙醇酒糟
fruit juices gelatin 果汁明胶
green liquor 绿液
glucose 葡萄糖
glycerin 甘油
grape juice 葡萄汁
green liquor 绿液
kraft liquor 牛皮纸浆
methanol stillage 甲醇废液
phosphoric acid 磷酸
potassium salts 钾盐
red liquor 媒染剂红液
slop 污油水
sodium salts solvents 钠盐溶剂
separation 分离
sewage gas 沼气
steepwater 浸泡水
sugars 糖
sulfates 硫酸盐
sulfites 亚硫酸盐
sulfides 硫化物
syrups 糖浆
tomato juice 番茄汁

urea　尿素
waste lubrication oil　废弃润滑油
waste water　废水
whey　乳清

Check Your Understanding

Exercise 1　Translate the following phrases into Chinese.

brewer's malt
coffee extract
sodium salts
solvents separation
citrus juice
ethanol stillage
whey
green liquor
kraft liquor
calcium salts
grape juice
copper sulfate
steep water
stick water
stillage
caustic soda
distiller slop
red liquor
collagen
gelatin

Exercise 2　Give brief answers to the following questions.

(1) What does ASME stand for?

(2) What are the product lines of Alaqua?

(3) What do Alaqua's evaporators run on?

(4) What is the function of Alaqua's evaporators?

(5) What does overall performance include?

(6) What do food products include?

Exercise 3 Translate the following sentences into chinese.

(1) Alaqua custom designs and manufactures a variety of lines of single and multiple effect evaporators.

(2) Alaqua custom designs and manutactures to ASME code and 3A sanitary standards.

(3) You can run Alaqua's evaporators on steam with high thermal economy, on hot oil or on compressors.

(4) Alaqua's evaporators will improve your overall performance by concentrating and crystallizing organic and inorganic chemicals.

阅读 C

Distillation

The separation operation called distillation utilizes vapor and liquid phases at essentially the same temperature and pressure for the coexisting zones. Various kinds of device such as dumped or ordered packings and plates or trays are used to bring the two phases into intimate contact. Trays are stacked one above the other and enclosed in a cylindrical shell to form a column. Packings are also generally contained in a cylindrical shell between hold-down and support plates.

1. Continuous Distillation

The feed material, which is to be separated into fractions, is introduced at one or more points along the column shell. Because of the difference in gravity between vapor and liquid phases, liquid runs down the column, cascading from tray to tray, while vapor flows up the column, contacting liquid at each tray.

Liquid reaching the bottom of the column is partially vaporized in a heated reboiler to provide boil-up, which is sent back up the column. The remainder of the bottom liquid is withdrawn as bottoms, or bottom products. Vapor reaching top of the column is cooled and condensed to liquid in the overhead condenser. Part of this liquid is returned to the column as reflux to provide liquid overflow. The remainder of the overhead stream is withdrawn as distillate, or overhead product.

This overall flow pattern in a distillation column provides countercurrent contacting of vapor and liquid streams on all the trays through the column. Vapor and liquid phases on a given tray approach thermal, pressure, and composition equilibrium to an extent dependent upon the efficiency of the contacting tray.

The lighter (lower-boiling) components tend to concentrate in the vapor phase, while the heavier (higher-boiling) components tend toward the liquid phase. The result is a vapor phase that becomes richer in light components as it passes up the column and a liquid phase that becomes richer in heavy components as it cascaded downward. The overall separation achieved between the distillate and the bottoms depends primarily on the relative volatility of the components, the number of contacting trays, and the ratio of the liquid-phase flow rate to the vapor-phase flow rate. If the feed is introduced at one point along the column shell, the column is divided into an upper section, which is often called the rectifying section, and a lower section, which is often referred to as the stripping section. These terms become rather indefinite in multiple-feed columns and columns from which a product sidestream is withdrawn somewhere along the column length in addition to the two end-product streams.

2. Batch Distillations

The simplest form of batch sill consists of a heated vessel (pot or boiler), a condenser, and one or more receiving tanks. No trays or packing are provided. Feed is charged into the vessel and brought to boiling. Vapors are condensed and collected in a receiver. No reflux is returned. The rate of vaporization is sometimes controlled to prevent bumping the charge and to avoid overloading the condenser, but other controls are minimum. This process is often referred to as Raleigh distillation.

The simple batch still provides only one theoretical plate of separation. Its use is usually restricted to preliminary work in which products will be held for additional separation at a later time, when most of the volatile component must be removed from the batch before it is processed further, or for similar noncritical separations.

To obtain products with a narrow composition range, a rectifying batch still is used that consists of a pot (or reboiler), a rectifying column, a condenser, some means of splitting off a portion of the condensed vapor (distillate) as reflux, and one or more receivers. Temperature of the distillate is controlled in order to return the reflux at or near the column temperature to permit a true indication of reflux quantity and improve column operation.

In operation, a batch of liquid is charged to the pot and the system is first brought to steady state under total reflux. A portion of the overhead condensate is then continuously withdrawn in accordance with the established reflux policy. Cuts are made by switching to alternate receivers, at which time operating conditions may be altered. The entire column operates as an enriching section. As time proceeds, composition of the material being distilled becomes less rich in the more volatile components, and distillation of a cut is stopped when accumulated distillate attains the desired average composition.

Check Your Understanding.

Exercise 1 Translate the following phrases into English or Chinese.

回流	in accordance with...
随意倾倒填料	batch distillations
成分平衡	boil-up
多组分物料流	multiple-feed

Exercise 2 Fill in the blanks with the correct words.

(1) Packings are generally contained in a cylindrical shell between ________ and________.

(2) Pressure, and composition equilibrium to an extent dependent upon ________.

(3) The simplest form of batch sill consists of____________________.

Exercise 3 Give brief answers to the following questions.

(1) What is the separation operation called distillation?

(2) Why do we need to control temperature of the distillate?

(3) When does distillation of a cut is stopped?

Exercise 4　Translate the following sentences into English.

(1) 平衡的程度取决于它们与塔板接触的效率。

(2) 简单精馏的应用通常局限于初步的分离工作，它的产品一般要留下待以后再分离。

(3) 气化速度有时要控制，以免出现液沫夹带或塔顶冷凝器超负荷的情况。

(4) 随着时间的推移，蒸馏物料的组成中挥发性较高的组分含量越来越低。

6.4 化 工 设 备

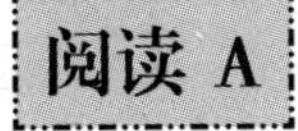

Hot New Products

1. Peristaltic Pumps Handles Abrasives and Corrosives

Vector series peristaltic pumps are designed to handle abrasive, corro-sive or high viscosity fluids commonly found in industrial processing. Roller-design and patented hose guide allow for powerful pumping action, without breaking up delicate emulsions or causing excessive frothing of dissolved gases. Pumps can reliably start, stop and continuously pump fluids that are pasty, pulpy, plastic or just plain thick at a wide range of pressures and flows. They are ideal for applications with aggressive fluids that require metering. Capacities are up to 60 PSI and 50 GPM. Pumps will reliably pump fluids that include solids up to 1/4 diameter.

(Instrument: Wanner Engrg., Inc., Hydra-Cell Pump Division, 1204 Chestnut Ave.,

Minneapolis, MN 55403.)

2. Depressing Pumps (Ideal for chemical applications)

These pumps ensure contamination-free transfer of abrasive and aggressive fluids such as acids, dyes and alcohol among others. In the 620CC and 800 Series, pumped product remains within smooth pump tube throughout entire pumping cycle, thereby eliminating any risk of dilution or cross contamination. Both pumps effectively handle high-pressure process applications. Pumps offer excellent metering performance, high vacuum lift, dry priming capability and no possibility of back-flow.

(Instrument: Watson Marlow Bredel, 37 Technology Park, Wilmington, MA 01887.)

3. Sensors

Booth 1977 Model 389VP and 399VP sensors are simpler to install, service and replace, which results in time and cost savings. Model 389VP is housed in a molded Tefzel(r) body with viton O-rings, making each sensor virtually indestructible and chemically resistant. Model 399VP disposable pH/ORP sensor is housed in a molded Tefzel body with 1" NMPT threads suit-able for insertion, submersion or flow-through installation.

(Instrument: Fisher-Rosemount Systems, Inc., 8301 Cameron Rd., Austin, TX 78754.)

4. Dry Screw Vacuum Pump

Booth 1441 Twister VSB is a cool, dry, environmentally friendly solution for solvent recovery and chemical processing applications. The pump uses a dual-pitch screw design. Advantages include lower operating temperatures, reduced power usage, clean compact design and an exclusive shaft seal design. These advantages are supplemented by addition of a high capacity booster blower for large throughput and deeper vacuums. This is an ideal system for processes that cannot tolerate even a hint of contamination or where environmental issues are a concern.

(Instrument: Rietschle Inc., 7222 Parkway Dr., Hanover, MD 21076.)

Check Your Understanding

Exercise 1 Mark the following statements with T(true) or F (false) according to the passage.

(1) Depressing pumps are designed to handle abrasives and corrosives. ( )

(2) High viscosity fluids are usually found in industrial processing. ( )

(3) Peristaltic pumps are ideal for applications with aggressive fluids that require metering. ( )

(4) Peristaltic pumps ensure contamination-free transfer of abrasive and aggressive fluids. ( )

(5) Dilution or cross contamination can be avoided by keeping pumped product within the smooth pump tube. ( )

(6) Depressing pumps have no possibility of back-flow. ( )

(7) Booth 1977 sensors can save time and cost. (　)

(8) Model 389VP sensors can make each censor hardly indestructible and chemically resistant. (　)

Exercise 2　Give brief answers to the following questions.

(1) Is it reliable for peristaltic pumps to pump fluids including solids up to 1/4" diameter?

(2) Can high-pressure process applications be effectively handled by both peristaltic pumps and depressing pumps?

(3) Where is Model 399VP disposable pH/ORP sensor housed?

(4) What is the advantage of 1" NMPT threads?

Exercise 3　Match the items listed in the following two columns.

(1) solvent recovery	a. 一点儿污染
(2) chemical processing	b. 流动状态
(3) lower operating temperature	c. 溶剂回收
(4) O-rings	d. 高真空提升
(5) chemically resistant	e. 交叉污染
(6) a hint of contamination	f. 化学处理过程
(7) flow-through installation	g. 压轮设计
(8) high vacuum lift	h. 运行温度低
(9) cross contamination	i. 圆环形
(10) roller-design	j. 耐化学腐蚀

Exercise 4 Translate the following phrases into English or Chinese.

English	Chinese
(1) ________________	微乳液
(2) excessive frothing	________________
(3) ________________	强腐蚀流体
(4) excellent metering performance	________________
(5) ________________	自启动功能
(6) exclusive shaft seal design	________________
(7) ________________	双齿距螺旋设计
(8) reduced power usage	________________
(9) ________________	紧凑、清洁的设计
(10) high capacity booster blower	________________

Exercise 5 complete the following sentences by translating the Chinese given in the brackets.

(1) Vector series peristaltic pumps________________________________(专门用来处理摩擦力大、腐蚀性强的流体). (be designed to)

(2) Peristaltic pumps can continuously pump fluids________________________(在较大压力和流动范围内).(at a wide range of)

(3) Depressing pumps________________________(是化工应用的理想设备). (be ideal for)

(4) Depressing pumps can______________________(消除稀释或交叉感染). (eliminate)

(5) Model 389VP censors______________________(易于安装、维修和拆换). (simpler to)

(6) Twister VSB__________________________(是一种理想的加工过程系统). (an ideal system)

阅读 B

Heat Exchangers

Heat exchangers arc so important and so widely used in the process industries that their design has been highly developed. Standards devised and accepted by the Tubular Exchanger Manufacture. Association (TEMA) are available covering in detail materials, methods of construction, technique of design, and dimensions for exchangers.

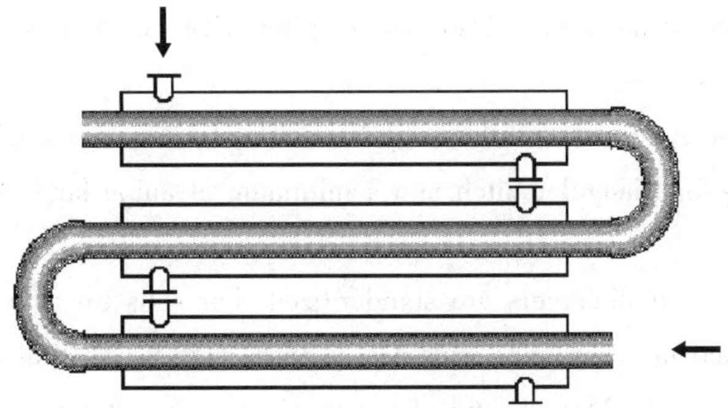

Figure 6.1　Double-pipe heat exchanger

图6.1　双管热交换器

The simple double-pipe exchanger shown in Figure 6.1 is inadequate for flow rates that cannot readily be handled in a few tubes. If several double pipes are used in parallel, the weight of metal required for the outer tubes becomes So large that the shell-and-tube construction, such as that shown in Figure 6.2, where one shell serves for many tubes, is more economical. This exchanger, because it has one shell-side pass and one tube-side pass, is a 1-1 exchanger.

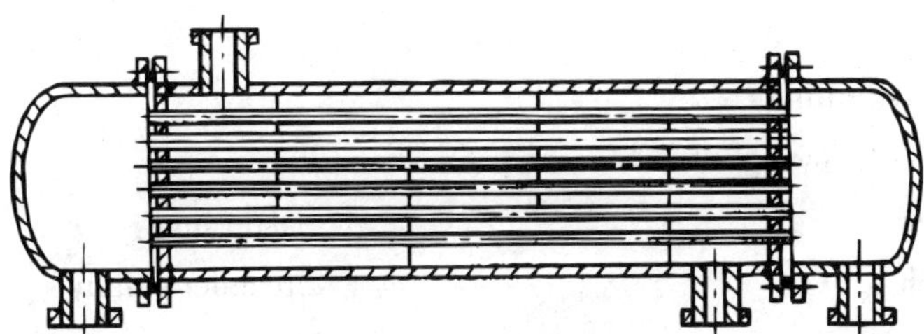

Figure 6.2　Single-pass counterflow heat exchanger

图6.2　单程逆流换热器

In an exchanger the shell-side and tube-side heat-transfer coefficients are of comparable importance, and both must be large if a satisfactory overall coefficient is to be attained. The velocity and turbulence of the shell-side liquid are as important as those of the tube-side fluid. To promote cross flow and raise the average velocity of the shell-side fluid, baffles are installed in the shell. In the construction shown in Figure 6.2, the baffles a consist of circular disks of her metal with one side cut away. Common practice is to cut away a segment having a height equal to one-fourth the inside diameter of the shell. Such baffles are called 25 percent baffles. The baffles are perforated to receive the tubes. To minimize leakage, the clearances between baffles and shell and tubes should be small.

Tubes and tube sheets standard lengths of tubes for heat-exchanger construction are 8, 12, 16, and 20 feet. Tubes are arranged on triangular or square pitch. Unless the shell side tends to foul badly, triangular pitch is used, because more heat-transfer area can be packed into a shell of given diameter than in square pitch. Tubes in triangular pitch cannot be cleaned by running a brush between the rows, because no space exists for cleaning lanes. Square pitch allows cleaning of the outside of the tubes. Also, square pitch gives a lower shell- side pressure drop than triangular pitch.

TEMA standards specify a minimum center-to-center distance 1.25 times the outside diameter of the tubes for triangular pitch and a minimum cleaning lane of 0.25 inch for square pitch .

Shell and baffle shell diameters are standardized. For sells up to and including 23 inch. The diameters are fixed in accordance with American Society for Testing and Materials (ASTM) pipe standards. For sizes of 25 inch and above the inside diameter is specified to the nearest inch. These shells are constructed of rolled plate. Minimum shell thicknesses are also specified.

The distance between baffles (center to center) is the baffle pitch, or baffle spacing. It should not be less than one-fifth the diameter of the shell or more than the inside diameter of the shell.

Tubes are usually attached to the tube sheets by grooving the holes circumferentially and rolling the tube ends into the holes by means of a rotating tapered mandrel, which stresses the metal of the tube beyond the elastic limit , so the metal flows into the grooves. In high-pressure exchangers, the tube are welded or brazed to the tube sheet after rolling.

Check Your Understanding

Exercise 1 Put the following into Chinese.

heat exchanger
triangular pitch
square pitch
guide rod
elastic limit
circumferentially
high-pressure exchangers
one-fifth

Exercise 2 Put the following into English.

套管式换热器
传热系数
管板
壳体直径
管子
折流挡板
加气器
安装管
角槽
并联

Exercise 3 Give brief answers to the following questions.

(1) How many parts does the heat exchanger consist of?

(2) What's the baffle? What's the roles of it?

(3) Why does square spacing provide lower shell side pressure drop than triangular spacing?

(4) How to minimize leakage?

Exercise 4 Translate the following passage into Chinese.

In an exchanger the shell-side and tube-side heat-transfer coefficients are of comparable importance, and both must be large if a satisfactory overall coefficient is to be attained. The velocity and turbulence of the shell-side liquid are as important as those of the tube-side fluid. To promote cross flow and raise the average velocity of the shell-side

阅读 C

Reactor Types

1. Stirred Tank Reactor

A **batch stirred tank reactor**(Figure 6.3) is the simplest type of reactor. It is composed of a reactor and a mixer such as a stirrer, a turbine wing or a propeller. The batch stirred tank reactor is illustrated in Figure 6.3.

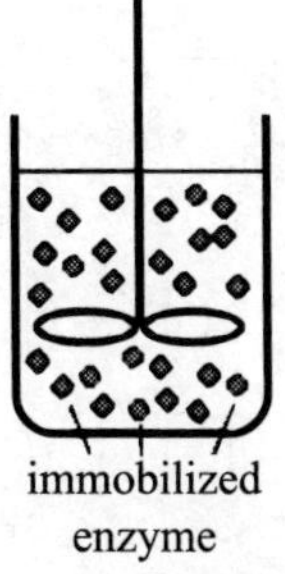

Figure 6.3 Batch stirred tank reactor
图6.3 间歇搅拌釜反应器

This reactor is useful for substrate solutions of high viscosity and for immobilized enzymes with relatively low activity. However, a problem that arises is that an immobilized enzyme tends to decompose upon physical stirring. The batch system is generally suitable for the production of rather small amounts of chemicals.

A continuous stirred tank reactor is shown in Figure 6.4.

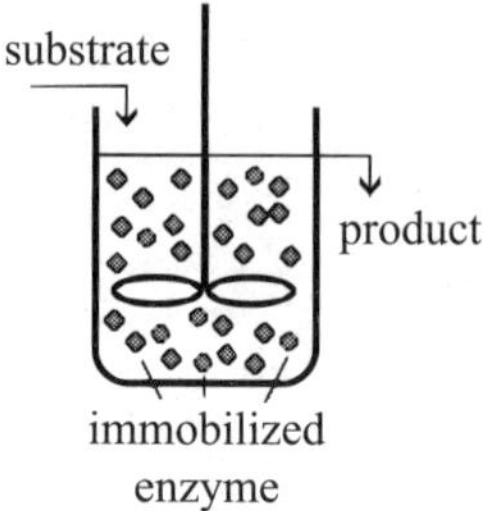

Figure 6.4 Continuous stirred tank reactor
图6.4 连续搅拌釜反应器

The continuous stirred tank reactor is more efficient than a batch stirred tank reactor but the equipment is slightly more complicated.

（摘自 http://www.rpi.edu/dept/chem-eng/Biotech-Environ/IMMOB/stirredt.htm.）

2. Tubular Reactor

Tubular reactors are generally used for gaseous reactions, but are also suitable for some liquid-phase reactions.

If high heat-transfer rates are required, small-diameter tubes are used to increase the surface area to volume ratio. Several tubes may be arranged in parallel, connected to a manifold or fitted into a tube sheet in a similar arrangement to a shell and tube heat exchanger. For high-temperature reactions the tubes may be arranged in a furnace.

3. Fluidized Bed Reactor

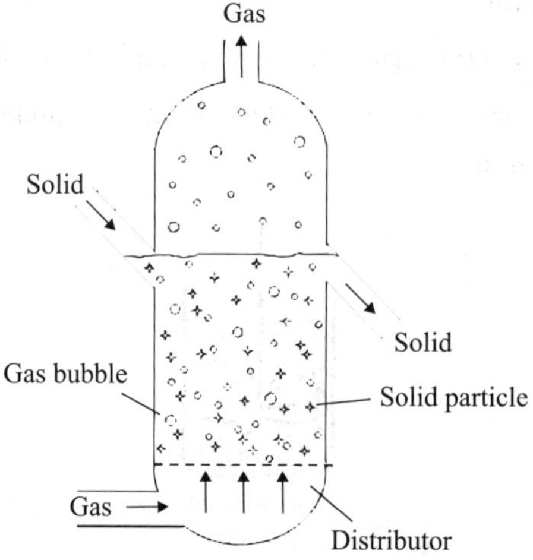

Figure 6.5 Fludized bed reactor
图6.5 流化床反应器

A fluidized bed reactor (FBR) is a type of reactor device that can be used to carry out a variety of multiphase chemical reactions. In this type of reactor, a fluid (gas or liquid) is passed through a granular solid material (usually a catalyst possibly shaped as tiny spheres) at high enough velocities to suspend the solid and cause it to behave as though it were a fluid (Figure 6.5). This process, known as fluidization, imparts many important advantages to the FBR. As a result, the fluidized bed reactor is now used in many industrial applications.

(1) Basic Principles

The solid substrate (the catalytic material upon which chemical species react) material in the fluidized bed reactor is typically supported by a porous plate, known as a distributor. The fluid is then forced through the distributor up through the solid material. At lower fluid velocities, the solids remain in place as the fluid passes through the voids in the material. This is known as a packed bed reactor. As the fluid velocity is increased, the reactor will reach a stage where the force of the fluid on the solids is enough to balance the weight of the solid material. This stage is known as incipient fluidization and occurs at this minimum fluidization velocity. Once this minimum velocity is surpassed, the contents of the reactor bed begin to expand and swirl around much like an agitated tank or boiling pot of water. The reactor is now a fluidized bed. Depending on the operating conditions and properties of solid phase various flow regimes can be observed in this reactor.

(2) Advantages

The increase in fluidized bed reactor use in today's industrial world is largely due to the inherent advantages of the technology.

• Uniform particle mixing: Due to the intrinsic fluid-like behavior of the solid material, fluidized beds do not experience poor mixing as in packed beds. This complete mixing allows for a uniform product that can often be hard to achieve in other reactor designs. The elimination of radial and axial concentration gradients also allows for better fluid-solid contact, which is essential for reaction efficiency and quality.

• Uniform temperature gradients: Many chemical reactions produce or require the addition of heat. Local hot or cold spots within the reaction bed, often a problem in packed beds, are avoided in a fluidized situation such as an FBR. In other reactor types, these local temperature differences, especially hotspots, can result in product degradation. Thus FBRs are well suited to exothermic reactions. Researchers have also learned that the bed-to-surface heat transfer coefficients for FBRs are high.

• Ability to operate reactor in Continuous State: The fluidized bed nature of these reactors allows for the ability to continuously withdraw product and introduce new reactants into the reaction vessel. Operating at a continuous process state allows manufacturers to produce their various products more efficiently due to the removal of startup conditions in batch processes.

(3) Disadvantages

As in any design, the fluidized bed reactor does have it draw-backs, which any reactor designer must take into consideration.

• Increased reactor vessel size: Because of the expansion of the bed materials in the reactor, a larger vessel is often required than that for a packed bed reactor. This larger vessel means that more must be spent on initial startup costs.

• Pumping requirements and pressure drop: The requirement for the fluid to suspend the solid material necessitates that a higher fluid velocity is attained in the reactor. In order to achieve this, more pumping power and thus higher energy costs are needed. In addition, the pressure drop associated with deep beds also requires additional pumping power.

• Particle entrainment: The high gas velocities present in this style of reactor often result in fine particles becoming entrained in the fluid. These captured particles are then carried out of the reactor with the fluid, where they must be separated. This can be a very difficult and expensive problem to address depending on the design and function of the reactor. This may often continue to be a problem even with other entrainment reducing technologies.

• Lack of current understanding: Current understanding of the actual behavior of the materials in a fluidized bed is rather limited. It is very difficult to predict and calculate the complex mass and heat flows within the bed. Due to this lack of understanding, a pilot plant for new processes is required. Even with pilot plants, the scale-up can be very difficult and may not reflect what was experienced in the pilot trial.

• Erosion of internal components: The fluid-like behavior of the fine solid particles within the bed eventually results in the wear of the reactor vessel. This can require expensive maintenance and upkeep for the reaction vessel and pipes.

(摘自 http://en.wikipedia.org/wiki/Fluidized_bed_reactor.)

4. Packed Bed Reactor

There are two basic types of packed-bed reactor: those in which the solid is a reactant, and those in which the solid is a catalyst. Many examples of the first type can be found in the extractive metallurgical industries.

In the chemical process industries the designer will normally be concerned with the second type: catalytic reactors. Industrial packed-bed catalytic reactors range in size from small tubes, a few centimeters diameter to large diameter packed beds. Packed-bed reactors are used for gas and gas-liquid reactions. Heat-transfer rates in large diameter packed beds are poor and where high heat-transfer rates are required fluidized beds should be considered.

Check Your Understanding

Exercise 1 Translate the following sentences into English or Chines.

间歇搅拌槽式反应器　　A continuous stirred tank

管状反应器	Uniform Temperature Gradients
流化床反应器	Particle Entrainment
填充床反应器	The high gas velocities

Exercise 2　Translate the following sentences into English.

(1) 它由反应器、搅拌器和涡轮机翼或螺旋桨等混合器组成。

(2) 管式反应器一般用于气相反应,但也适用于某些液相反应。

(3) 目前对流化床中材料实际行为的理解相当有限。

(4) 填充床反应器有两种基本类型: 固体为反应物的反应器和固体为催化剂的反应器。

(5) 在化学过程工业中,设计者通常关心的是第二种类型:催化反应器。工业填充床催化反应器的尺寸范围从小管,直径几厘米到大直径填充床。

(6) 填充床反应器有两种基本类型:一种是固体作为反应物,另一种是固体作为催化剂。

(7) 大直径填料床的热传导率较差,当需要较高的热传导率时,应考虑流化床。

Exercise 3 Translate the following passage into Chinese.

(1) This reactor is useful for substrate solutions of high viscosity and for immobilized enzymes with relatively low activity. However, a problem that arises is that an immobilized enzyme tends to decompose upon physical stirring. The batch system is generally suitable for the production of rather small amounts of chemicals.

(2) However, a problem that arises is that an immobilized enzyme tends to decompose upon physical stirring. The batch system is generally suitable for the production of rather small amounts of chemicals.

(3) Tubular reactors are generally used for gaseous reactions, but are also suitable for some liquid-phase reactions.

(4) Increased reactor vessel size: Because of the expansion of the bed materials in the reactor, a larger vessel is often required than that for a packed bed reactor. This larger vessel means that more must be spent on initial startup costs.

(5) Industrial packed-bed catalytic reactors range in size from small tubes, a few centimeters diameter to large diameter packed beds. Packed-bed reactors are used for gas and gas-liquid reactions.

6.5 化工安全

Safety in the Laboratory

You will carry out many laboratory activities in advanced chemical topics. While no human activity is completely risk free, if you use common sense and a bit of chemical sense, you will encounter few problems. Chemical sense is an extension of common sense. Sensible laboratory conduct won't happen by memorizing a list of rules, any more than a perfect score on a written driver's test ensures an excellent driving record. The true test of chemical sense is your actual conduct in the laboratory. The following safety rules apply to all laboratory activities. For your personal safety and that of your classmates, make following these guidelines second nature in the laboratory. If you understand the reasons behind them, these safety rules will be easy to remember and to follow.

Rules of laboratory conduct:

(1) Perform laboratory work only when your teacher is present. Unauthorized or unsupervised laboratory experimenting is not allowed.

(2) Your concern for safety should begin even before the first activity. Always read and think about each laboratory assignment before starting.

(3) Know the location and use of all safety equipment in your laboratory. These should include the safety shower, eye wash, first-aid kit, fire extinguisher, and blanket.

(4) Wear a laboratory coat or apron and protective glasses or goggles for all laboratory work. Wear shoes (rather than sandals) and tie back loose hair.

(5) Clear your bench top of all unnecessary materials such as books and clothing before starting your work.

(6) Check chemical labels twice to make sure you have the correct substance. Some chemical formulas and names differ by only a letter or number. Pay attention to the hazard classifications shown on the label.

(7) You may be asked to transfer some laboratory chemicals from a common bottle or jar to your own test tube or beaker. Do not return any excess material to its original container unless authorized by your teacher. Do not pipette by mouth-ever.

(8) Avoid unnecessary movement and talk in the laboratory.

(9) Never taste laboratory materials. Gum, food, or drinks should not be brought into the

laboratory. If you are instructed to smell something, do so by fanning some of the vapor toward your nose. Do not place your nose near the opening of the container.

(10) Never look directly down into a test tube; view the contents from the side. Never point the open end of a test toward yourself or your neighbor.

(11) Any laboratory accident, however small, should be reported immediately to your teacher.

(12) In case of a chemical spill on your skin or clothing rinse the affected area with plenty of water. If the eyes are affected water-washing must begin immediately and continue for 10 to 15 minutes or until professional assistance is obtained.

(13) Minor skin burns should be placed under cold, running water.

(14) When discarding used chemicals, carefully follow the instructions provided.

(15) Return equipment, chemicals, aprons, and protective glasses to their designated locations.

(16) Before leaving the laboratory, ensure that gas lines and water faucets are shut off.

(17) If in doubt, ask!

(摘自 http://www.files.chem.vt.edu/RVGS/ACT/lab/safety_rules.html.)

Check Your Understanding

Exercise 1 Translate the expressions into English or Chinese.

许多实验活动	laboratory work
化学常识	safety shower
安全设备	designated locations
防护眼镜	gas lines

Exercise 2 Translate the following sentences into English.

(1) 未经授权或无人监管的实验室是不允许进入的。

(2) 了解实验室所有安全设备的位置和用途。

(3) 避免在实验室内进行不必要的活动和交谈。

（4）丢弃用过的化学品时，请仔细遵循提供的说明。

（5）如果你的皮肤或衣服被化学物质溅到，应用大量的水冲洗受影响的区域。如果眼睛受到影响，必须立即用水冲洗，并持续10~15分钟直到获得专业帮助。

Exercise 3 Translate the following passage into Chinese.

（1）For your personal safety and that of your classmates, make following these guidelines second nature in the laboratory. If you understand the reasons behind them, these safety rules will be easy to remember and to follow.

（2）Know the location and use of all safety equipment in your laboratory. These should include the safety shower, eye wash, first-aid kit, fire extinguisher, and blanket.

（3）Check chemical labels twice to make sure you have the correct substance. Some chemical formulas and names differ by only a letter or number. Pay attention to the hazard classifications shown on the label.

（4）Never taste laboratory materials. Gum, food, or drinks should not be brought into the laboratory. If you are instructed to smell something, do so by fanning some of the vapor toward your nose. Do not place your nose near the opening of the container.

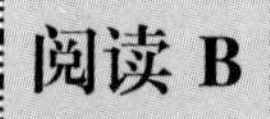

Hazards of Dangerous Chemicals

Many chemicals which are in common use have safety and/or health hazards associated with them. However, forethought and careful working can minimize these hazards. The following are typical hazards of some dangerous chemicals and main precautions.

1. Corrosive Chemicals

Corrosive chemicals are those substances which by direct chemical action are injurious to the body or destroy metal. Their ability to cause injury depends on the nature of the corrosive, the temperature and concentration, and the site and duration of exposure.

The following guidelines should be observed:

(1) Suitable respiratory protective equipment should be available and used when required.

(2) Use effective fume cupboards.

(3) Know emergency procedures. In particular, know how to obtain running water with which to irrigate an injury.

2. Cancer Causing Chemicals

Some chemicals are thought to be able to produce cancer in man at levels that could be reached in working, unless proper care is taken. Other chemicals are merely suspected of this, sometimes on the basis of research using animals. Nevertheless, all such substances must be treated with caution.

Carcinogenic activity is often found in the following classes of compounds:

Polycyclic aromatics, bicyclic or polycyclic aromatic amines, nitro and nitroso compounds, nitroso phenols, epoxides, reactive alkyl halides and 1-haloethers, aziridines, 2-propiolactone organic sulphates, diazomethane, nitrogen mustards, other alkylating agents, nitrosamines.

Users of carcinogens should be aware of the risk in handling them. A sense of proportion is in order. The well publicized chemical tragedies have mostly arisen from prolonged industrial exposure or from people actually eating the offending substances. Information on particular substances can be obtained from safety & environmental protection services.

3. Flammability

Flammable substances cause many fires and explosions every year, which are avoidable if simple precautions are taken.

Not all concentrations of a flammable vapor in air are flammable. For each vapor, there is a concentration below which and above which propagation of a flame will not occur. These are known as the lower and upper explosive limits, and vary widely for carbon monoxide: 12.5%~74%; carbon disulphide: 1%~50%; Benzene: 1.5%~8%.

As these substances are found ubiquitously, it is imperative that they are stored safely: Keep stocks of flammable liquids to an absolute minimum. Safe storage of flammable liquids for immediate use can be provided by ventilated steel cabinets with overlapping hinged doors. Drip trays are provided to contain the contents of the largest bottle in the store. Mark cabinets conspicuously as to their contents and situate them in the laboratory at a point remote from the exits.

Under no circumstances will smoking be allowed near any operation where flammable liquids are being handled or used, or near any storeroom or cabinet where flammable liquids are kept. It goes without saying that ignition sources such as gas burners, sparking electrical contacts etc. must be excluded from areas where highly flammable liquids are stored or used.

4. Explosions

The main causes of explosion include compounds which readily detonate some compounds, such as acetylides, peroxides and nitro compounds, decompose explosively on little provocation. Such compounds should only be prepared if it is considered essential for a research provocation. Even then, limit batch size severely. Operate behind a safety screen, wearing heavy gloves, overalls, a face visor and hearing protection. Place the apparatus so that no one could be injured if there was an explosion in the system.

Check Your Understanding.

Exercise 1　Translate the expressions into English or Chinese.

安全和健康危害	simple precautions
腐蚀性化学品	explosive limits
温度和浓度	flammable liquids
多环芳香化合物	ignition sources
易燃物	causes of explosion

Exercise 2　Translate the following sentences into English.

(1) 许多常用的化学品在安全和健康方面都会产生危害。

(2) 腐蚀性化学品是对人体有伤害或可破坏金属的物质。

(3) 了解紧急程序,采取有效急救措施。

(4) 在空气中,并不是任意浓度的易燃蒸气都可以燃烧。

(5) 必须安全保存危险化学品。

Exercise 3 Make use of the following key words to write down an essay.

离开实验室时,为了减少污染带来的危险,要执行下列措施:如果戴手套,先用冷水洗之并脱去,再用冷水和肥皂洗手,温水会使化学试剂进入皮肤,化学试剂挥发还会刺激眼睛;脱掉实验服,如果污染严重,将其放在塑料袋内并密封,然后立即洗手。(提示:suitably, penetrate, operation, laboratory, glove, carry out, wear, contamination, in the order, irritate, remove, skin, volatilize, plastic。)

Exercise 4 Translate the following passage into Chinese.

Some chemicals are thought to be able to produce cancer in man at levels that could be reached in working, unless proper care is taken. Other chemicals are merely suspected of this, sometimes on the basis of research using animals. Nevertheless, all such substances must be treated with caution.

阅读 C

Chemical Process Safety Information

Complete and accurate written information concerning process chemicals, process technology, and process equipment is essential to an effective process safety management (PSM) program.

1. Chemicals in the Process

The information about the chemicals needs to be comprehensive enough for an accurate assessment of the fire and explosion characteristics, reactivity hazards, the safety and health hazards to workers, and the corrosion and erosion effects on the process equipment and monitoring tools.

2. Technology of the Process

Process technology information uses diagrams that will help users understand the process better.

The block flow diagram is a simplified diagram. It is used to show the major process equipment and interconnecting process flow lines and flow rates, temperatures, and pressures when necessary for clarity.

Process flow diagrams are more complex and show all main flow streams including valves, vessels, feed and product lines, heat exchangers and points of pressure and temperature control. Also, information on construction materials, pump capacities and pressure heads, compressor horsepower, and vessel design pressures and temperatures are shown when necessary for clarity. In addition, process flow diagrams usually show major components of control loops along with key utilities.

3. Equipment in the Process

Piping and instrument diagrams (P&IDs) may be the more appropriate type diagrams to show some of the above details. The P&IDs are used to describe the relationships between equipment and instrumentation.

4. Employee Involvement

Employers are to consult with their employees and their representatives regarding their efforts in developing and implementing the process safety management program elements and hazard assessments. They also train and educate their employees and inform affected employees of the findings from incident investigations. Many employers already have established methods to keep employees informed about relevant safety and health issues and may be able to adapt these practices and procedures to meet their obligations under PSM.

5. Process Hazard Analysis

A process hazard analysis (PHA) is one of the most important elements of PSM program. It provides information that will assist employers and employees in making decisions for improving safety.

A PHA analyzes potential causes and consequences of fires, explosions, releases of toxic or flammable chemicals, and major spills of hazardous chemicals. The PHA focuses on equipment, instrumentation, utilities, human actions (routine and non-routine), and external factors that might affect the process.

6. Operating Procedures

Operating procedures describe tasks to be performed, data to be recorded, operating conditions to be maintained, samples to be collected, and safety and health precautions to be taken. The procedures need to be technically accurate, understandable to employees, and revised periodically to ensure that they reflect current operations. The process safety information ensures that the operating procedures and practices are consistent with the known hazards of the chemicals in the process and that the operating parameters are correct. Operating procedures should be reviewed by engineering staff and operating personnel to ensure their accuracy and that they provide practical instructions on how to actually carry out job duties safely. Also the employer must certify annually that the operating procedures are current and accurate.

Check Your Understanding

Exercise 1 Translate the expressions into English or Chinese.

(1) 程序框图
(2) 流速
(3) 工艺流程图
(4) 对工人造成的安全和健康危害
(5) 书面信息
(6) pump capacities and pressure heads
(7) piping and instrument diagrams
(8) process hazard analysis
(9) compressor horsepower
(10) heat exchangers

Exercise 2 Translate the following sentences into English.

(1) 危险化学品的准确信息对于实施有效的安全管理十分必要。

(2) 流程图通常显示控制回路的主要部件。

(3) 管道设备布置图说明了设备和仪表的关系。

(4) 过程危险性分析是过程安全管理最重要的要素之一。

(5) 操作程序描述了要执行的工作任务。

Exercise 3　Translate the following sentences into English.

化工厂是生产化工产品的加工厂。化工厂的工作就是通过化学或生物反应生产新材料。化工厂在生产过程中要使用专门的设备、单元和技术。其他的工厂,比如制药厂、炼油厂、生物化工厂和废水处理厂会使用许多和化工厂类似的技术。实际上,有些人把它们视为化工厂。

Exercise 4　Translate the following passage into Chinese.

Piping and instrument diagrams (P&IDs) may be the more appropriate type diagrams to show some of the above details. The P&IDs are used to describe the relationships between equipment and instrumentation.

第3篇

专业英语写作

第7章 化学化工英语论文的结构

What is a Scientific Paper?

A paper is an organized deion of hypotheses, data and conclusions, intended to instruct the reader. Papers are a central part of research. If your research does not generate papers, it might just as well not have been done. "Interesting and unpublished" is equivalent to "non-existent".

Realize that your objective in research is to formulate and test hypotheses, to draw conclusions from these tests, and to teach these conclusions to others. Your objective is not to "collect data".

A paper is not just an archival device for storing a completed research program, it is also a structure for planning your research in progress. If you clearly understand the purpose and form of a paper, it can be immensely useful to you in organizing and conducting your research. A good outline for the paper is also a good plan for the research program. You should write and rewrite these plans/outlines throughout the course of the research. At the beginning, you will have mostly plan; at the end, mostly outline. The continuous effort to understand, analyze, summarize, and reformulate hypotheses on paper will be immensely more efficient for you than a process in which you collect data and only start to organize them when their collection is "complete".

化学化工论文属于科技论文范畴,是科学研究成果的书面表达形式,论文需要准确、清晰地表达研究内容和获得的成果,提高论文的可读性,合理安排论文的结构,使每一部分内容详实、论述完整、结构逻辑严谨。

7.1 论文的分类

论文既可按写作目的进行分类,也可按论文内容进行分类。

7.1.1 按写作目的分类

论文根据写作目的可分为期刊论文、学位论文和会议论文。

(1) 期刊论文是供学术期刊发表的,它在科技论文中占有最大比例。大部分期刊论文对字数提出了上限,大致为6000~8000字。不同级别的期刊对论文的深度、广度要求不同,总体来说,论文从选题到内容,都需要创新性,且在理论上具有较强的指导意义或在应用中有较高的实用价值。

(2) 学位论文分为学士论文、硕士论文、博士论文。每篇学位论文都需要有几个科技创新点,特别是博士论文,一般要求至少有4个创新点。

(3) 会议论文供参加学术会议使用。各种会议论文水平相差很大,有时其深度、广度有限,但会议论文的优点是可以更快地反映当前研究的现状,参会者或读者可由此了解国内外的研究热点及研究水平。

7.1.2 按论文内容分类

科技论文根据内容可分为实验型、理论型和综述型3种。

(1) 实验型论文的重点在于通过科学实验提供事实,其中包括新方法或新工艺,这种论文常有大量数据,以及由数据得出的结论。

(2) 理论型论文可以完全不涉及实验(例如,数学类论文),而是引用文献的实验结果,由此提出新规律、新模型、新计算方法。在同一篇论文中也常见实验和理论部分同时存在。

(3) 综述型论文是对某一课题前期工作的评述,总结某一方法、技术。这种论文常常需要调研大量的文献,并对这些文献做出客观中肯的评价,指出其优缺点,指出今后的研究方向。因此,综述型论文对同方向的研究者有指导作用,可作为今后工作的基础。

7.2 科技论文的结构

英文科技论文最常见结构是IMRAD结构,即包括Introduction、Methods、Results and Discussions等部分。

IMRAD结构的论文清晰明了且逻辑性强,首先阐述研究的课题和研究目的,再描述研究的方法、试验手段和材料,最后对结果和结论进行详细讨论。一般来说,一篇完整、规范的学术论文由以下各部分构成:

(1) 标题(Title)。

(2) 作者和联系方式(Author Affiliation)。

(3) 摘要(Abstract)。

(4) 关键词(Key words)。

(5) 目录(Table of contents)。

(6) 术语表(Nomenclature)。

(7) 引言(Introduction)。

(8) 方法(Method)。

(9) 结果(Result)。

(10) 讨论(Discussion)。

(11) 结论(Conclusion)。

(12) 致谢(Acknowledgement)。

(13) 参考文献(Reference)。

(14) 附录(Appendix)。

7.2.1　标题

论文标题原则上要用最少的词表述最核心的内容。科技论文的标题是论文内容的高度概括,拟定标题时需要注意以下几点。

1. 标题长度

标题长度不宜过长,大多为8~12个单词(中文一般要求不超过20个字)。国际标准化组织规定,每个标题不要超过12个单词,且除通用的缩写词或特殊符号外,标题内不能使用缩写词和特殊符号。如果12个单词不足以概括全文的内容,可使用副标题加以补充说明。

2. 标题用词

为了便于检索,标题中所用的词尽量使用表达全文内容的关键词,因此会有多个论文的关键词出现在标题中。标题通常由名词短语构成,即由一个或多个名词加上其前置定语或后置定语构成,因此标题中使用最多的是名词,其次是介词、形容词等,若出现动词,一般是现在分词、过去分词或动名词形式。

撰写标题时,一个简单有效的方法是找出一个可以反映论文核心内容的主题词,将其扩展成名词短语,使之包含论文的关键信息,要注意词语间的修饰要恰当。

3. 标题的结构

标题主要有3种结构:名词词组、介词词组、名词词组+介词词组,一般不使用不定式短语,也不使用从句。

(1) 名词词组:由名词及其修饰语构成。

例如:Phase transition and electriacl properties.

(2) 介词短语(使用较少),一般使用的介词为on,表示对……的研究。

例如:On the Distribution of Sound in a Corridor.

(3) 名词词组(名词)+介词词组,介词短语一般用来修饰名词或名词词组,从而限定该研究课题的范围。

例如:New ferroelectrics of complex composition.

Phase transition and electriacl properties of new lead-free $(Na_{0.5}K_{0.5})NbO_3$-$Ba(Ti_{0.95}Hf_{0.05})O_3$ solid solution ceramics.

标题的书写形式有以下3种,其中前两种使用较广泛:

(1) 除标题首个单词第一个字母、专有名词大写外,其余均小写,要求这种格式的刊物有*Remote Sensing*、西安电子科技大学学报等。

(2) 标题首个单词第一个字母、实词及多于5个字母的介词、连词第一个字母大写,其余小写,要求这种格式的刊物有西北工业大学学报、西安交通大学学报等。

(3) 全部字母均大写。

例如:

A study of the phase equilibria between fluorothene and some solvents such as dibutyl phthalate and chlorotrifluoro ethylene at elevated temperature.

关于在高温下聚三氯氟乙烯塑料与邻苯二甲酸二丁酯和一氯三氟乙烯等溶剂之间的相平衡的研究。

这个标题的缺点包括:

(1) 标题太长,不醒目,难以引证。

(2) "A study of" 不增加标题含义,只增加长度,属于废词,可以去掉。

(3) 使用了一般的词,不简练,如"some solvents such as"都增加了标题长度,标题之所以冗长,常常是由于使用专业词汇少,而使用一般词汇多。

再如:

A New Procedure for the Analysis of Steel Constituents.

一种分析钢组成的新方法。

这个标题中专业性词汇很少,只有"Analysis"和"steel"两个,如改写为"Spectrophotometric Determination of Iron and Chromium in steel",其中专业词就有5个。

【练习7.1】 讨论下列标题是否合适,并指出如何修改。

(1) Electromagnetic Fields Have Harmful Effects on Humans.

(2) How to Use Water Resources for Irrigation in Semiarid Land.

(3) Diamond is used for Electronic Devices.

(4) Water Quality Can Be Protected Through the Successful Integration of Research and Education.

7.2.2 作者和联系方式

论文的署名表明作者享有著作权,获得相应的荣誉和利益,但也表示文责自负,即所有作者均有义务对其发表的研究成果的科学性和真实性负责。论文署名还便于作者与同行或读者进行研讨和联系,因此作者有必要提供尽可能详细的联系方式,如工作单位和通讯地址,包括邮政编码等。

作者姓名以及所在单位编排在题名下面,作者多于一个时,按所做贡献排。当作者单位不一致时,也要进行标注。例如:

Processing and Morphology of (111) $BaTiO_3$ Crystal Platelets by a Two-Step Molten Salt Method

Danya Lv, Ruzhong Zuo*, and Shi Su

Institute of Electro Ceramics and Devices, School of Materials Science and Engineering, Hefei University of Technology, Hefei 230009, China

"*"代表通讯联系人(一般放在最后),电子邮件、电话、传真号码等以脚注形式标注。

7.2.3　摘要

摘要是论文的缩影,是对论文的简单描述。摘要的作用是为读者提供关于文献内容的足够信息,即论文所包含的主要概念和所讨论的主要问题,可使读者从摘要中获得作者的主要研究活动、研究方法和主要结果及结论。摘要可以帮助读者判断此论文是否有助于自己的研究工作,是否有必要获取全文。一篇好的摘要应该具备以下几方面的特点:

(1) Motivation: Why do we care about the problem and the results?

(2) Problem statement: What problem is the paper trying to solve and what is the scope of the work?

(3) Approach: What was done to solve the problem?

(4) Results: What is the answer to the problem?

(5) Conclusions: What implications does the answer imply?

例如:

Abstract: In this paper, we propose a new implementation of multiplication—free binary arithmetic coding by use of table lookup which reduce the complexity of hardware design. The scheme can be shown to have better approximation of the used probability model and to have minor performance degradation.

摘要:本文提出了利用查找来实现二值算术编码,避免乘除法运算,可以简化硬件设计,该算法具有较小的概率逼近误差,因此性能退化较小。

关于摘要的撰写,我们将在第8章具体阐述。

7.2.4　关键词

为便于读者选读及检索文献,每篇论文在摘要后应给出3~8个关键词(中文一般是3~5个),其作用是反映主题。

关键词的选取要注意使读者能据此大致判断论文的研究内容,并且一般按照研究目的—研究方法—研究结果的顺序标引关键词。

7.2.5 引言

引言的主要任务是向读者勾勒出全文的基本内容和轮廓。它可以包括以下五项内容中的全部或其中几项:

(1) 介绍某研究领域的背景、意义、发展状况、目前水平等。

(2) 对相关领域的文献进行回顾和综述,包括前人的研究成果、已经解决的问题,并适当加以评价或比较。

(3) 提出前人尚未解决的问题,也可提出新问题,以及解决这些新问题的新方法、新思路,从而引出自己研究课题的动机和意义。

(4) 说明自己研究课题的目的。

(5) 概括论文的主要内容,或勾勒其大体轮廓。

可以将引言分为三到四个层次来安排:

第一层:① Introducing the general research area including its background, importance and present level of development...

② Reviewing previous research in this areas...

第二层:Indicating the problem that has not been solved by previous research, raising a relevant question...

第三层:Specifying the purpose of your research...

第四层:① Announcing your major findings...

② Outline the contents of your paper...

引言中各层次所占篇幅可以有很大差别,其中第一个层次会占去大部分篇幅,研究背景和目前研究状况这两部分会有比较详细的介绍,而研究目的这部分会比较简短。

比较简短的论文,引言也可以相对比较简短:

第一层:Introducing the importance of the research area and reviewing previous research...

第二层:Indicating the problem that has not been solved by previous research, raising a relevant question...

第三层:Specifying the purpose of your research...

引言的各个层次不仅有其各自任务和目的,在语言上也有各自特点,下面分别做介绍。

1. 引言开头的写法

引言的最主要目的是告诉读者论文所涉及的研究领域及意义是什么,即要回答以下问题:

① What is the subject of the research?

② What is the importance of this subject?

③ How is the research going at present?

④ In what way is it important, interesting and worthy studying?

⑤ What problem does the research solve?

因此,在写引言开头时,首先,关键词往往出现在第一句话中,迅速地将主题告诉读者,然后再简单介绍该研究的意义。例如:

Forecast of the tracks of hurricanes have improved steadily over the past three decades. These improvements have brought about a significant decline in loss of life.

其次,引言开头句子的谓语动词一般是一般现在时或现在完成时,形成一些常用的句型。

(1) 研究主题+be

例如:**Spot welding is** the most widely used joining method in the automobile manufacturing industry.

(2) 研究主题+has become

例如:**Forest decline has become** a favorite topic for environmental studies.

(3) 研究主题+被动语态

例如:**Air pollution has been extensively studied** in recent years.

(4) there has been growing interest in/concern about +研究主题

例如:In the 1990s, **there has been growing interest in** the development of electric vehicles in response to the public demand for cleaner air.

2. 引言文献综述的写法

文献综述是作者对他人在某研究领域所做的工作和研究成果的总结及评述,包括引用他人有代表性的观点或理论、发明或发现、解决问题的方法等。

在援引他人的研究成果时,必须标注出处。标注出处时有两种方法,一种是著者+出版年,另一种是顺序编码。

例1:Hanson et al. (1976) presented this new method.[或This new method was presented by Hanson et al. (1976).]

例2:A new method was presented by Hanson et al.[1](或et al.[1-3]).

如果需要多处引用同一个作者或同一篇文章,那么需要使用一些连接手段使上、下文衔接。例如:

① The author goes on to say that...

② The article further states that...

③ The author also argues that...

3. 研究动机与研究目的的写法

介绍研究动机可从两个角度入手:一是指出前人尚未解决的问题,二是说明解决这一问题的重要意义。在指出前人尚未解决的问题时有一些常用句型,下面进行详细介绍。

(1) 用表示否定意义的词(如little、few、no或none of)+名词作主语:

① Little information/work/research...

② Few studies/investigations/researchers...

③ No studies/data...

④ None of these studies/findings...

例如:There have been **few specific reports** in the literature of oak and hickory decline.

（2）用表示对照的句型：

① The research has tended to focus on... rather than on...

② Although considerable research has been devoted to... rather less attention has been paid to...

（3）提出问题或假设：

① However, it remains unclear whether...

② If these results could be confirmed, they would provide strong evidence for...

提出问题后下一步应指出本研究的目的和内容，可使用以下句型：

（1）描述研究目的最简单的句型是用this paper、this study、this project、this research等词作主语，后面用investigate、discuss、examine等动词作谓语，宾语为研究内容，即

This paper This study This project The present study This research Our project This survey This thesis	+	concerns tests investigates reports discusses describes explains calculates examines analyses proposes demonstrates measures

（2）把论文本身当作强调的内容，通常采用一般现在时。例如：

① The purpose of this paper is to discuss...

② The aim of this report is to evaluate...

③ The objective of the present paper is to examine...

（3）把研究活动作为主语，通常采用被动语态。例如：

① This research is designed to determine...

② This study is designed to measure...

③ Our project aim to calculate...

4. 引言结尾的写法

写引言时可以把研究目的作为引言的结尾，也可以简单介绍文章的结构和各部分的主要内容，使读者了解文章的轮廓。

介绍文章结构时注意避免使用同一个句型结构，例如：

① Section 1 describes...

② Section 2 analyses...

③ Section 3 discusses...

可以使用如下结构：

① ...Section 2 defines...

② ...is then presented in section 3.

③ followed by...

④ ...is evaluated in section 4.

⑤ while section 5 discusses...

⑥ Finally, we find that...

⑦ in section 6, there is...

⑧ ...is given...

7.2.6　论文主体

科技论文按内容可分为实验型、理论型和综述型，下面分别介绍各类型论文的论文主体。

1. 实验型论文的论文主体

实验型论文阐述的核心内容是实验，以及进一步对实验结果进行定性或定量的讨论。实验型论文包括的主要内容如下：

(1) 实验用原料和方法(Materials and methods)或实验(Experimental)

应详细说明实验所使用的仪器设备(种类、型号等)，说明实验和观察的步骤和方法。论述应能使本领域的科学家在必要时可以重复作者的工作并能判断出作者结论的可靠性和有效性。给出的信息不必过分详细，描述的重点是新的工作和新的方法，对于已发表的一般标准技术、方法或设备，只需扼要介绍或举出参考文献即可。

例如：

2. Experimental

(1-x)NKN-xBA (0≤ x ≤ 0.1) powders were prepared from the constituent oxides and carbonates by a conventional solid-state reaction method. Appropriate mixtures of Na_2CO_3 (99.8%), K_2CO_3 (99.0%), Nb_2O_5 (99.5%), Bi_2O_3 (99.5%) and Al_2O_3 (99.5%) were ball mixed in ethanol for 12 h. After mixing, the slurry was dried, crushed and then calcined in a lidded alumina crucible. The mixing and calcination were repeated two times for homogenization. A further milling process of the calcined powders together with 0.5 wt% PVA binder was finished in a nylon jar with ZrO_2 balls for 24 h. These powders were subsequently pressed into disks under a uniaxial pressure of 50 MPa. The powder compacts were sintered for 3 h in air. The microstructure was observed by a scanning electron microscope (SEM, JEOL6301F, Tokyo, Japan). The crystal structure was examined by means of an X-ray diffractometer (XRD, Rigaku, Japan). For electrical measurements, specimens were polished and painted with a silver paste on two major surfaces. The dielectric properties were measured by a LCR meter (HP4284A, Hewlett-Packard, USA) equipped with a programmable temperature box. Electric poling was done by immersing samples into silicone oil and then applying a dc field of 3 kV/mm for 20 min. After 24 h, the piezoelectric constant d_{33} was measured by a quasi-static Berlincourt meter (YE2730, Sinocera, Yangzhou, China). The planar electromechanical coupling factor kp was determined by a resonance-antiresonance method with an impedance analyzer (HP4192A, Hewlett-Packard, USA).

(2) 实验结果和讨论(Results and discussion)

实验所得到的结果和数据,是完成并决定论文质量的关键和推论的依据,所以应该准确充分地表达出来,论述要简明、客观、真实,不可有任何夸张虚假,要能使读者容易理解并信服,提出的资料应主要是新发现的和有代表性的,而不是冗长而重复的资料,不是所有研究得到的资料都必须或都可以写入论文,研究工作所积累的资料必须通过整理、加工、改造等数理统计处理和技术处理,并分析判断其中存在的关系,得出某些见解,形成概念和判断,然后用文字说明或绘图、列表或用符号形式表达实验结果。例如:

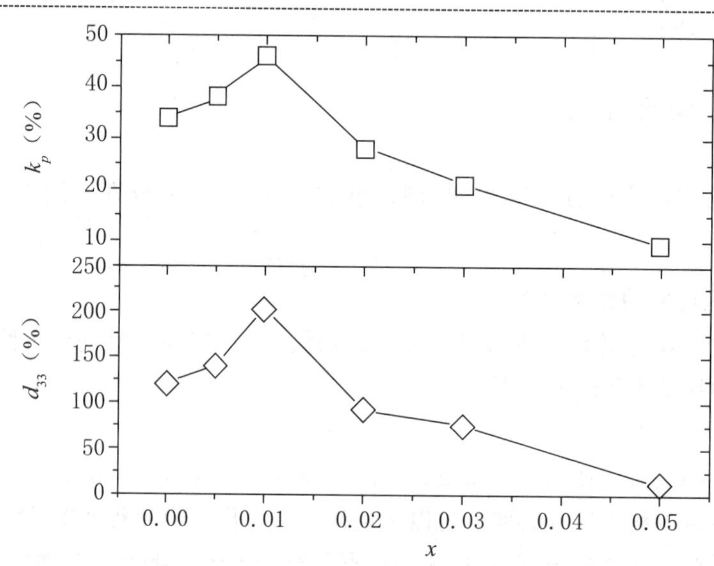

Figure 5 Compositional dependence of piezoelectric constant d_{33} and electromechanical coupling factor k_p of poled(1-x)NKN-xBA ceramics.

The piezoelectric and electromechanical properties of poled $(1-x)$ NKN-xBA ceramics with compositions near the MPB are shown in Figure 5 The addition of a small amount of BA greatly enhances the piezoelectric activity of NKN ceramics. The best properties with a d_{33} of 202 pC/N and a k_p of 46% appear in the composition with x = 0.01. With further increasing or decreasing x, the properties drop down rapidly. It is evident that a strong compositional dependence of piezoelectric properties exists near the MPB. It is evident that the MPB plays a crucible role in the enhancement of piezoelectric properties.

(3) 结论(Conclusion)

结论部分主要写出将实验得到的资料、数据经过总结、归纳、判断、抽象、推论而最后得到的主要结论,文字一定要简明,结论一定要精炼、可靠。当然论文中也不一定要把此部分单独列出来写。除了总结关键内容之外,还需对研究成果的应用前景进行展望。例如:

4. Summary

Lead free $(1-x)$NKN-xBA piezoelectric ceramics have been manufactured by ordinary sintering. The results indicate that the addition of BA significantly influences the sintering, microstructure, phase transition and electrical properties of NKN ceramics. The identification of phase transitional behavior confirms the formation of a MPB between orthorhombic and tetragonal ferroelectric phases in the composition range of $0.005 \leqslant x \leqslant 0.01$. The enhanced ferroelectric and piezoelectric properties with a d_{33} of 202 pC/N, a k_p of 46%, a P_r of 23.6 μC/cm² and a Curie temperature of 372 ℃ appear in the composition with x = 0.01. The results indicate that (1−x)NKN-xBA ceramics could be a potential candidate for lead free piezoelectric materials.

2. 理论型论文的论文主体

理论型论文是以理论研究为主体的学术论文,在进行理论阐述时要注意逻辑推理。理论阐述包括论证的理论依据、对其所做的假设与合理性进行的论证、对于分析方法所做的说明。在写作时,应注意区别哪些是已知的、哪些是作者第一次提出的、哪些是经过作者改进的,等等,这些都应作详细说明。

总之,理论阐述的要点是:假设、前提条件、分析的对象、所引用的数据及其可靠性、适用的理论或新模型的提出、分析方法、计算过程、新理论或模型的验证、导出的结论。

3. 综述型论文的论文主体

综述型论文的内容一般包括问题的提出、发展历史的介绍、现状分析、未来发展趋势和建议。

综述型论文的价值在于,作者通过对已有文献的综合和归纳,分析某学术领域或某一方面研究工作的发展历史和现状,指出该领域科学活动的发展方向,提出具有科学性、创造性和前瞻性的研究课题。

下面是论文主体中的一些常用句型。

(1)描述实验方法的常用句型

描述实验方法与过程的句型通常采用被动语态的过去式,因为在描述实验过程或方法时,句子中的主题或中心是实验材料、场地和方法本身,表达“做了什么”“怎么做的”之意,而不是表达“谁做了什么”,因此,在描述方法时,常将实验材料等作为主语,谓语动词自然要用被动语态。例如:

① ...the sections **were rinsed** to remove adherent material and dried...

② The adventitious roots(不定根)from each section **were removed**...

③ Five agents **were selected** at random and **asked** to collect a water sample for the instructor.

(2)理论分析的常用句型

在正文部分的假设、理论分析、数理模型的建立以及计算过程中,常常要推导并描述公式,作者应了解和掌握公式的规范表达和描述公式的基本句型。科技论文理论分析部分常用到数字与逻辑关系的表述,这类表述是不受时间影响的普遍事实,所以表示理论分析的句型常使用一般现在时态。例如:

① Substituting M in/into N, we obtain/have/get...

② Substituting M in/into N gives/yields/results in...

③ **Let us now consider the case** the dynamic link parameters are unknown.

④ **Suppose that** X_t is a solution of the SDE(1).

⑤ **If** m=1.2, then we have the following equations...

⑥ **Given that** m=1.2, we obtain...

⑦ **The relationship** between m and n is as follows...

描述公式时需注意以下几点：每个公式占一行，居中排；有两个或两个以上公式时，要有编号，编号加上圆括号，列在右端；行文中，将公式看作一个单词处理，若是一句话的结束，需加句号，否则根据情况选用合适的标点；若公式中的符号是第一次出现的，要加以解释，用where或in which引导。公式中的符号也作为一个单词处理，所以符号后谓语动词用单数形式。行文中涉及某一公式时，要用“Equation...”或“Equ...”表示。例如：

① The current in the wire is calculated using

$$V=IR,$$

where(in which) V is voltage, I is current, and R is resistance.

② By eliminating l between Equations(1) and (2), we have...

③ Let f be a continuous function...

④ Suppose that g is a constant...

⑤ When x is close to 0, then y is close to 1.

(3) 图表文字说明中的常用句型

为了清楚地描述研究结果，作者可以将结果用图表表示。注意若是表，表题要写在表上面，若是图，图题要写在图下面。图表中的内容要用文字来说明。在文字说明中，如果提及图表，表格用“Table+序号”表示，如“Table 1 shows...”。其他各种类型的图都可以用“Figure+序号”表示，如“from figure 2 we learn...”。“Figure”这个词可以用缩写“Fig.”，但必须前后统一。

文字说明中的动词要用一般现在时态，选用的动词要慎重，因为不同的动词表示研究结果的可靠性不同，同时也表示作者不同的判断、推理和态度。例如：

① **Table 1 shows/provides** details of the test results.

② **Figure 4.2 gives** the results of the experiment.

③ Both of these questions received very high ratings (**see Table 1**).

上述句型也可使用被动语态，例如：

① As shown **in Table 3**...

② As can be seen from the data **in Table 1**...

③ As described **in Figure 3.2**...

(4) 介绍研究成果的常用句型

作者对他的研究成果的评论或说明通常包含下列内容之一。

① 根据本人的研究结果做出推论。例如：

These results suggest that untrained octopuses(章鱼)can learn a task more quickly by observing the behavior of another octopus than by reward-and-punishment methods.

② 作者解释研究结果或说明产生研究结果的原因。例如：

These findings are understandable because the initial annealing(退火)temperature dictates(规定)the state of conformational(构象的) structures.

③ 作者对此次研究结果与其他研究者曾发生的结果作比较，例如提出自己的结果是否与其他研究者的结果一致。例如：

These results agree with Gerner's analysis, in that Q_{max} varies inversely with length and to the third power of the pipe width.

④ 作者对自己的研究方法或技术的性能与其他研究者的方法或技术的性能进行比较。例如：

The recognition rate of our system **is significantly higher than** that reported for Token's system.

⑤ 作者指出自己的理论模型是否与实验数据符合。例如：

The measured temperatures along the heat pipe **are all highly consistent with** the predictions of the theoretical model.

当评论的内容是对研究结果可能的证明时，句子的主要动词之前通常加上may或can等一般现在时态的情态动词。例如：

① The layer structural or some other mixed complex material **may be the most suitable refrigerants** for the Ericesson magnetic refrigerator.

② **One reason of this advantage may be that** the Hank visual programming language can avoid some of the syntactic problems associated with textual programming languages.

③ **A possible explanation for this is that...**

④ **These results agree well with** the findings of Cassdio, et al.

当作者指出由自己的理论模型所得到的预测与实验数据之间的吻合程度时，通常使用一般现在时态，说明模型预测与实验数据是否一致不受时间影响。例如：

① The data **confirm closely** the prediction of the model.

② The theoretical model **fits** the experimental data **well.**

③ The theoretical model **agrees well with** the experimental data.

④ The experimental measurements **are very close to** the predicted values.

(5) 讨论部分的常用句型

在讨论部分中，作者通常对研究结果进行概述、分析、解释等，因为陈述的是作者的见解和结论，所以句型多用一般现在时态。

其中，概述结果通常表示如下：

① **These results provide substantial evidence for** the original assumptions.

② **These experimental results support** the original hypothesis that...

③ These results **contradict** the original hypothesis.

④ **The present results are consistent with** those reported in our earlier work.

当表示研究的局限性时，常用表述如下：

① It should be noted that **this study has examined only...**

② The findings of **this study are restricted to...**

③ **The limitations of this study** are clear...

④ The result of the study **cannot be taken as evidence for...**

⑤ Unfortunately, **we are unable to determine from this data...**

7.2.7 致谢

致谢对象主要包括两个部分:在研究过程中给予你帮助的个人和单位以及在研究经费上给予帮助的外单位或基金会。常用句式有:

① The author is indebted to... for... and to... for...

② The author would like to express their appreciation to...

③ In particular we would like to acknowledge the contribution of...

④ The author wishes to thank...for

⑤ This work was supported by...

⑥ The authors are grateful to... for their fruitful discussion.

例如:

Acknowledgements

This work was financially supported by HFUT RenCai Foundation (No. 103-035006) and a special Program for Excellence Selection "R&D of Novel Lead Free Piezoelectric Ceramics" (No.103-035034), an open fund of State Key Laboratory of New Ceramics and Fine Processing and Nippon Sheet Glass Foundation for Materials Science and Engineering.

7.2.8 参考文献

参考文献的具体写法有两种情况:按作者姓名字母顺序排列(alphabetical list of references)和按引用顺序排列(numbered list of references)。

在具体编排时,有的出版物要求采用将作者名字写在前面、姓写在后面的格式,而有的出版物要求姓写在前、名写在后。遇到多作者情况时,即可以将各作者的姓名一一写出,也可以只写出第一作者,其余用"et al."表示,有的出版物则要求写出前三名作者,其余用"et al."表示。不同出版物有不同要求,要有针对性地参阅有关书刊上的实例。但不管有什么要求,都必须注意两点:

(1) 只应列入重要的已发表的文章。与文章关系不大的不要列进去。尚未发表的资料、尚在印刷中的文章、学位论文、文摘等一般也不要列进去。

(2) 引证的文章应该是自己阅读过的,而不要转引自其他文章,否则可能会造成差错。

例如:

References

[1] Y. Saito, H. Takao, T. Tani, T. Nonoyama, K. Takatori, T. Homma, T. Nagaya, M. Nakamura, Nature 432 (2004) 84.

[2] L. Egerton, D.M. Dillon, J. Am. Ceram. Soc. 42 (1959) 438.

[3] R.Z. Zuo, J. Rodel, R.Z. Chen, L.T. Li, J. Am. Ceram. Soc. 89 (2006, 2010).

...

材料阅读

Outlines

1. The Reason for Outlines

I emphasize the central place of an outline in writing papers, preparing seminars, and planning research. I especially believe that for you, and for me, it is most efficient to write papers from outlines. An outline is a written plan of the organization of a paper, including the data on which it rests. You should, in fact, think of an outline as a carefully organized and presented set of data, with attendant objectives, hypotheses and conclusions, rather than an outline of text.

An outline itself contains little text. If you and I can agree on the details of the outline (that is, on the data and organization), the supporting text can be assembled fairly easily. If we do not agree on the outline, any text is useless. Much of the time in writing a paper goes into the text; most of the thought goes into the organization of the data and into the analysis. It can be relatively efficient to go through several (even many) cycles of an outline before beginning to write text; writing many versions of the full text of a paper is slow.

All the writing that I do—papers, reports, proposals (and, of course, slides for seminars)—I do from outlines. I urge you to learn how to use them as well.

2. How Should You Construct an Outline?

The classical approach is to start with a blank piece of paper, and write down, in any order, all important ideas that occur to you concerning the paper. Ask yourself the obvious questions: "Why did I do this work?" "What does it mean?" "What hypothesis did I mean to test?" "What ones did I actually test?" "What were the results?" "Did the work yield a new method or compound? What?" "What measurements did I make?" "What compounds? How were they characterized?" Sketch possible equations, figures, and schemes. It is essential to try to get the major ideas written down. If you start the research to test one hypothesis, and decide, when you see what you have, that the data really seem to test some other hypothesis better, don' t worry. Write them both down, and pick the best combinations of hypotheses, objectives and data. Often the objectives of a paper when it is finished are different from those used to justify starting the work. Much of good science is opportunistic and revisionist.

When you have written down what you can, start with another piece of paper and try to organize the jumble of the first one. Sort all of your ideas into three major heaps.

(1) Introduction

Why did I do the work? What were the central motivations and hypotheses?

(2) Results and Discussion

What were the results? How were compounds made and characterized? What was measured?

(3) Conclusions

What does it all mean? What hypotheses were proved or disproved? What did I learn? Why does it make a difference?

Next, take each of these sections, and organize it on yet finer scale. Concentrate on organizing the data. Construct figures, tables, and schemes to present the data as clearly and compactly as possible. This process can be slow—I may sketch a figure 5~10 times in different ways, trying to decide how it is most clear (and looks best aesthetically).

Finally, put everything—outline of sections, tables, sketches of figures, equations—in good order.

When you are satisfied that you have included all the data (or that you know what additional data you intend to collect), and have a plausible organization, give the outline to me. Simply indicate where missing data will go, how you think (hypothesize) they will look, and how you will interpret them if your hypothesis is correct. I will take this outline, add my opinions, suggest changes, and return it to you. It usually takes 4~5 repeated attempts (often with additional experiments) to agree on an outline. When we have agreed, the data are usually in (or close to) final form (that is, the tables, figures, etc., in the outline will be the tables, figures...in the paper.)

You can then start writing, with some assurance that much of your prose will be used.

The key to efficient use of your and my time is that we start exchanging outlines and proposals as early in a project as possible. Do not, under any circumstances, wait until the collection of data is "complete" before starting to write an outline. No project is ever complete, and it saves enormous effort and much time to propose a plausible paper and outline as soon as you see the basic structure of a project. Even if we decide to do significant additional work before seriously organizing a paper, the effort of writing an outline will have helped to guide the research.

3. The Outline

What should an outline contain?

(1) Title

(2) Authors

(3) Abstract

Do not write an abstract. That can be done when the paper is complete.

(4) Introduction

The first paragraph or two should be written out completely. Pay particular attention to the opening sentence. Ideally, it should state concisely the objective of the work, and indicate why this objective is important.

In general, the Introduction should have these elements:

- The objectives of the work.
- The justification for these objectives.

Why is the work important?

- Background.

Who else has done what? How? What have we done previously?

- Guidance to the reader.

What should the reader watch for in the paper? What are the interesting high points? What strategy did we use?

- Summary conclusion.

What should the reader expect as conclusion? In advanced versions of the outline, you should also include all the sections that will go in the Experimental section (at this point, just as paragraph subheadings).

(5) Results and Discussion

The results and discussion are usually combined. This section should be organized according to major topics. The separate parts should have subheadings in boldface to make this organization clear, and to help the reader scan through the final text to find the parts that interest him or her. The following list includes examples of the phrases that might plausibly serve as section headings:

- Synthesis of alkane thiols.
- Characterization of monolayers.
- Absolute configuration of the vicinal diol unit.
- Hysteresis correlates with roughness of the surface.
- Dependence of the rate constant on temperature.
- The rate of self-Exchange decreases with the polarity of the solvent.

Try to make these section headings as specific and information-rich as possible. For example, the phrase "The Rate of Self-Exchange Decreases with The Polarity of The Solvent" is obviously longer than "Measurement of Rates," but much more useful to the reader. In general, try to cover the major common points:

- Synthesis of starting materials.
- Characterization of products.
- Methods of characterization.
- Methods of measurement.
- Results (rate constants, contact angles, whatever).

In the outline, do not write any significant amount of text, but get all the data in their proper place: any text should simply indicate what will go in that section.

- Section headings.
- Figures (with captions).

- Schemes (with captions and footnotes).
- Equations.
- Tables (correctly formatted).

Remember to think of a paper as a collection of experimental results, summarized as clearly and economically as possible in figures, tables, equations, and schemes. The text in the paper serves just to explain the data, and is secondary. The more information that can be compressed into tables, equations, etc., the shorter and more readable the paper will be.

(6) Conclusion

In the outline, summarize the conclusions of the paper as a list of short phrases or sentences. Do not repeat what is in the Results section, unless special emphasis is needed. The Conclusions section should be just that, and not a summary. It should add a new, higher level of analysis, and should indicate explicitly the significance of the work.

(7) Experimental

Include, in the correct order to correspond to the order in the results section, all of the paragraph subheadings of the experimental section.

4. In Summary

- Start writing possible outlines for papers early in a project. Do not wait until the "end". The end may never come.
- Organize the outline and the paper around easily assimilated data—tables, equations, figures, schemes-rather than around text.
- Organize in order of importance, not in chronological order. An important detail in writing paper concerns the weight to be given to topics. Neophytes often organize a paper in terms of chronology: that is, they recount their experimental program, starting with their cherished initial failures and leading up to a climactic successful finale. This approach is completely wrong. Start with the most important results, and put the secondary results later, if at all. The reader usually does not care how you arrived at your big results, only what they are. Shorter papers are easier to read than longer ones.

第8章　摘要的撰写

中国国家标准和美国国家标准都对论文摘要做了定义：

(1) 中国国家标准定义(CNS 13152, 1993)：一篇精确代表文献内容的简短文字，不多加阐释或评论，也不因撰写摘要的人不同而有差异。

(2) 美国国家标准定义(ANSI Z39.14, 1979)：一篇精确代表文献内容的简短文字。

依据上述定义，显然定义的摘要与论文中由作者所写的摘要略有不同，这种摘要以信息性或指示性的成分居多，往往非作者本人所写。在英文中，与摘要有关的英文单词有Abstract、Summary与Synopsis三种，严格而言，Abstract属摘录性质，用于在专业性数据库中查询；Summary则有“摘要”或“概要”的意思，是再次简明扼要地陈述全文较突出的发现或结论。Synopsis亦有“概要”及“大意”之意，是对全文的概略描述。

本章所述的摘要以Abstract为主，是对研究论文正文的精炼概括，便于读者在最短时间内了解全文内容，通常收录于相应学科的摘要检索类数据库或专刊内。其目的是为读者提供论文所包含的主要研究活动、研究方法、研究结果和结论。可以帮助读者判断这篇论文对自己的研究工作是否有用，是否有必要获取全文，为科研人员、科技情报人员及计算机检索提供方便。

8.1　摘要的种类和特点

8.1.1　摘要的种类

摘要按不同功能来划分，大致有以下 3 种类型：

(1) 报道性摘要。报道性摘要是文献的主题范围及内容梗概的简明摘要，相当于简介。报道性摘要一般用来反映论文的目的、方法、主要结果与结论，在有限的字数内向读者提供尽可能多的定性或定量的信息，充分反映该研究的创新之处。因此，学术性期刊(或论文集)多选用报道性摘要，篇幅以 300 字左右为宜。

(2) 指示性摘要。指示性摘要指明论文的论题、取得成果的性质和水平，其目的是使读者对该研究的主要内容(即作者做了什么工作)有一个轮廓性的了解。创新内容较少的论文可选用指示性摘要，一般适用于学术性期刊的简报、问题讨论等栏目以及技术性期刊等，篇幅以 100 字左右为宜。

(3) 报道-指示性摘要。报道-指示性摘要是指以报道性摘要的形式表述论文中价值最大的部分,其余部分则以指示性摘要的形式表现出来,篇幅以 100~200 字为宜。

一般地,向学术性期刊投稿,应选用报道性摘要形式;只有创新内容较少的论文,其摘要可写成报道-指示性或指示性摘要。

根据论文种类来划分,摘要主要有以下3种类型:

(1) 学术刊物上的摘要。常置于主体部分前,目的是让读者首先了解论文的内容,以便决定是否阅读全文。一般这种摘要是在全文完成之后才写的,字数一般在100~150字,但也有很长的摘要。摘要内容包括研究目的、研究方法、研究结果和主要讨论。

(2) 学术会议论文摘要。往往在会议召开几个月前撰写,交由会议论文评审委员会评阅,从而决定能否录用。这种摘要比第一种略为详细,长度在200~300字。开头有必要介绍研究课题的意义、目的、宗旨等。如果在撰写摘要时研究工作尚未完成,全部研究成果还未得到,那么应在方法、目的、宗旨、假设方面多花笔墨。

(3) 学位论文摘要。一般在400字左右(视所做工作而定,可长可短),根据需要分为几个段落。内容一般包括研究背景、意义、主旨和目的,基本理论依据、基本假设,研究方法,研究结果,主要创新点,简短讨论。要突出创新之处,应指出有何新的观点、见解或解决问题的新方法。

8.1.2 摘要的特点

1. 英文摘要的时态

英文摘要时态的运用也以简练为佳,多用一般现在时、一般过去时,少用现在完成时、过去完成时,进行时态和其他复合时态基本不用。具体内容,详见8.3.1小节。

2. 英文摘要的语态

采用何种语态,既要考虑摘要的特点,又要满足表达的需要。一篇摘要很短,尽量不要随便混用语态,更不要在一个句子里混用。

(1) 主动语态

现在主张摘要中谓语动词尽量采用主动语态的越来越多,皆因其有助于文字清晰、简洁及表达有力。例如:

In this paper, we propose a new implementation of multiplication—free binary arithmetic coding by use of table look up which reduce the complexity of hardware design.

(2) 被动语态

以前强调多用被动语态,理由是科技论文主要是说明事实经过的,至于那件事是谁做的,无需一一证明。事实上,在指示性摘要中,为强调动作承受者,还是采用被动语态为好。即使在报道性摘要中,有些情况下被动者无关紧要,也必须用要强调的事物做主语。例如:

This investigation was performed to see if the aluminum in pop cans and aluminum cookware enters the liquid they contain. It was hypothesized that aluminum does enter the liquids in aluminum cans and cookware.

3. 英文摘要的人称

以往摘要的首句多用第三人称 this paper...开头，现在倾向于采用更简洁的被动语态或原形动词开头。例如：to describe...、to study...、to investigate...、to assess...、to determine...等。行文中最好不用第一人称，以方便文摘刊物的编辑刊用。

4. 词汇

(1) 冠词

在冠词中定冠词the易被漏用。the用于表示整个群体、分类、时间、地名以外的独一无二的事物、形容词最高级等较易掌握，但用于特指时常被漏用。这里有个原则，即当我们用the 时，听者或读者已经确知我们所指的是什么。例如：

The author designed a new machine.

The machine is operated with solar energy.

由于现在缩略语越来越多，要注意区分a和an.

(2) 数词

避免用阿拉伯数字作首词，如："Three hundred Dendrolimus tabulaeformis larvae are collected..."中的"Three hundred"不要写成"300"。

(3) 单复数

一些名词单复数形式不易辨认，从而造成谓语形式出错。

8.2　摘要的结构和内容

8.2.1　摘要的结构

1. 学术期刊论文摘要

简短精练是学术期刊论文摘要的主要特点，只需简明扼要地将研究目的、方法、结果和讨论分别用1~2句话加以概括即可。至于研究背景和宗旨应在论文Introduction中详细介绍。这种摘要结构布局如下：

(1) Title.

(2) Author(s), address.

(3) Objectives, purposes, hypothesis...

(4) Methods, materials, procedures...

(5) Result, data, observations, discussion...

(6) Conclusion...

2. 学术会议论文摘要

学术会议论文摘要直接决定论文是否被录用，因此首先要简要说明研究背景、内容、范

围、价值和意义,在研究方法上也可以多花一些笔墨。这种摘要结构布局如下:

(1) Title.

(2) Author(s), address.

(3) Background, previous studies, present situation, problems that need to be solved...

(4) Objectives of this study, hypothesis...

(5) Methods, materials, procedures...

(6) Results, data, observations, discussion...

(7) Conclusions...

3. 学位论文摘要

学位论文摘要一般单独占一页,也可能占两页,装订在学位论文目录之前。学位论文摘要也要介绍研究背景、内容、目的、方法、结果等。但是学位论文摘要的不同之处在于它必须指出研究结果的独到之处或创新点。关于研究的内容也可稍加详细介绍。这种摘要结构布局如下:

(1) Title.

(2) Author(s), address.

(3) Background, problems that need to be solved, rationale for the present study...

(4) Objectives and scope of this study...

(5) Outline of the main contents and results...

(6) Conclusion...

8.2.2 摘要的内容

以上3种摘要具有共性,内容一般都包括:

(1) 目的(objectives, purposes):包括研究背景、范围、内容、要解决的问题及解决这一问题的重要性和意义。

(2) 方法(methods and materials):包括材料、手段和过程。

(3) 结果与简短讨论(results and discussions):包括数据和分析。

(4) 结论(conclusions):包括主要结论、研究的价值和意义等。

摘要内容的具体要求如下:

(1) 文摘要尽量简短。

① 取消不必要的字句,如: It is reported..., Extensive investigations show that..., The author discusses..., This paper concerned with...,等等;文摘开头的In this paper以及一些不必要的修饰词,如in detail、briefly、here、new、mainly也尽量不要。

② 论文第一句应避免与题目(Title)重复。

③ 对物理单位及一些通用词可以适当进行简化。

④ 取消或减少背景信息。

⑤ 限制文摘只表述新情况、新内容,过去的研究细节可以省略。

⑥ 不说无用的话,如“本文所谈的有关研究工作是对过去老工艺的一个极大的改进”“本工作首次实现了……”“经检索尚未发现与本文类似的文献”等词句切不可进入文摘。

⑦ 作者在文献中谈及的未来计划不纳入文摘。

(2) 文摘应包含正文的要点,将文章的主要内容写清楚,重点内容不能漏掉,比如实验研究的方法、设备、材料等,一定要给出结论。

(3) 文摘应有自我独立性和自明性。

(4) 不需要自我标榜研究成果。

(5) 文摘中不能出现图、表数据。

(6) 语句简洁,专业词汇准确。

8.3 英文摘要的时态和人称

8.3.1 英文摘要的时态

英文摘要时态的运用以简练为佳,多用一般现在时、一般过去时,少用现在完成时、过去完成时、进行时态和其他,复合时态基本不用。

1. 一般现在时

一般现在时用于说明研究目的、叙述研究内容、描述结果、得出结论、提出建议或讨论等。例如:

With the developments of embedded technology and popularization of the internet, embedded internet technology has become a new focus in the research of embedded system which is the most expeditious new technology.

The result shows (reveals)...

It is found that...

The conclusions are...

The author suggests...

涉及公认事实、自然规律、永恒真理等,当然也要用一般现在时。

2. 一般过去时

一般过去时用于叙述过去某一时刻(时段)的发现、某一研究过程(实验、观察、调查、仿真等过程)。例如:

At the same time, the author erected a kind of EICS, the stresses of which are the system safety checking and the implement of EIU.

需要指出的是,用一般过去时描述的发现、现象,往往是尚不能确认为自然规律、永恒真理的。

3. 现在完成时和过去完成时

完成时要少用，但并非不用。现在完成时把过去发生的或过去已完成的事情与现在联系起来，而过去完成时可用来表示过去某一时间以前已经完成的事情，或在一个过去事情完成之前就已完成的另一过去行为。例如：

Concrete has been studied for many years.

Man has not yet learned to store the solar energy.

8.3.2 英文摘要的人称

以往摘要的首句多用第三人称 This paper... 开头，现在倾向于采用更简洁的被动语态或原形动词(To describe，To study，To investigate，To assess，To determine)开头。例如：

The torrent classification model and the hazard zone mapping model are developed based on the geography information system.

【练习8.1】 将下面的英文摘要翻译成中文。

In this paper, a typical Li and Ta/Sb-modified alkaline niobate based ceramics prepared by conventional sintering and two-step sintering were investigated in terms of their phase transition behaviors and various properties. The phase structures of the ceramics sintered by conventional sintering shifted remarkably from a polymorphic phase transition (PPT) to a tetragonal symmetry with the increase of sintering temperature. However, the ceramics sintered by two-step sintering maintained PPT over a wide sintering temperature range. Similar to that, the various properties of the ceramics sintered by conventional sintering are strongly dependent on sintering temperature and the ceramics with good properties can only be obtained in a narrow sintering temperature range, while the ceramics with excellent properties were obtained by two-step sintering over a wide sintering temperature range. The results indicate that two-step sintering is an effective way to broaden the sintering temperature range of (K, Na)NbO_3-based ceramics.

【练习8.2】 将下面的中文摘要翻译成英文。

通过传统固相法，在1350 ℃条件下保温3 h，合成纯的高结晶度的四方$BaTiO_3$(BT)粉体。这种固相法合成的BT粉体将作为熔盐法合成片状$Ba_6Ti_{17}O_{40}$(B6T17)粉体的原料粉体。当热处理温度为1150 ℃，保温3 h时，B6T17形成了完整的片状结构。采用定向颗粒层XRD测试技术分析(OPL-XRD)，显示片状B6T17粉体沿(001)面择优生长。以B6T17为前驱体，制备的片状BT粉体具有最高的纵横比(直径为10~20 μm，厚度小于1.5 μm)。此外，OPL-XRD分析显示，BT沿(111)面择优生长。

材料阅读

How to Write an Abstract?

1. What is an Abstract?

An abstract is a summary of points of a writing presented in skeletal form, which highlights the important information covered in an article or a paper. It helps the reader to find out quickly whether the writing is of their interest or not, and whether they need to read the whole paper or not. Moreover, national or international conference organizers decide whether a participant is eligible or not usually by reviewing the abstract submitted. Therefore, writing a good abstract is of great importance for scholars and researchers.

2. Criteria of a Good Abstract

In terms of content, a good abstract states clearly the purpose of the writing, the method used, the findings, the originality, or the implication etc. EI, for example, requires a good abstract in English should answer the following questions:

(1) What do you want to do?

(2) How did you do it?

(3) What results did you get and what conclusion did you draw?

(4) What is original in your paper?

To meet the requirements, a good abstract should be accurate, concise, specific, self-contained, and coherent.

• Being accurate means that an abstract presents the information that actually occurs in the paper, and avoids vagueness in definition, word choice and elsewhere.

• Being concise means that an abstract includes only the most important ideals, findings, or implications, and avoids wordy expressions. There is a word limit for abstract of journals or periodicals, usually within 250 words. The length varies. One should comply with the specific requirements of the journal that he/she is going to publish his/her paper. However, an abstract should not be over-simplified; otherwise the readers cannot get sufficient information about the paper.

• Being specific means that an abstract should be concrete, and to the point rather than general and indirect so as to reduce vagueness and misunderstanding.

• Being self-contained means an abstract is complete in itself. It covers the important points in the paper, with formalized structure showing the topic, the supporting evidence, and the conclusion.

• Being coherent means that an abstract is logical and make sense. A good abstract uses tense correctly, and avoids too many changes between active voice and passive voice.

3. Types of Abstract

There are three major types of abstract: descriptive, informative, and the combination of the two.

(1) Descriptive (Indicative) Abstract

A descriptive or indicative abstract presents the general subject matter of the paper. It aims at introducing the research briefly, so it does not provide specifics of the research. It tends to be short.

(2) Informative (Informational) Abstract

An informative abstract highlights the findings and results. It provides some essential data of the research. Therefore, it tends to be specific and sometimes quantitative.

(3) Informational-Indicative Abstract

An informational-indicative abstract is a combination of the above two types. It includes not only the general information of the research but also the essential data of the finding and result.

An abstract is usually one paragraph of running text, summarizing the most important elements of a paper. No matter what type you choose, which depends, to certain extent, on the specific requirements of the journal, you should enable the readers to gasp the essential elements accurately and quickly.

4. Steps of Writing an Abstract

(1) Draft the abstract when you finish writing the paper, for you have a clearer picture of all your ideals, findings etc. expressed in the paper.

(2) Structure the abstract in almost the same order as your paper. Begin with a key sentence stating the subject of the research and indicating the objective of the paper; then continue with a brief summary of the research method, experiment, procedure, investigation, result, analysis, and discussion; finally end with summing up the conclusion.

Each abstract has its own features, due to the subject matter that the paper explores. Usually it follows the sequence of the original paper and highlights the essential point in it.

5. Difference between Abstract and Introduction

Some Chinese students have no idea about what an abstract is, so they simply translate the introduction of the paper into English and put it at the beginning as an abstract. In fact, they differ from one another in their purposes and contents. The following table 8.1 brief sketches the difference of the two.

Table 8.1　Differences between abstract and introduction

图8.1　摘要和简介之间的区别

Abstract	Introduction
The essence of the whole paper	Introduce the topic of the paper
Covers the following academic elements: • Background • Purpose and focus • Methods • Results • Conclusions • Recommendations	Covers the following academic elements: • Background • Scope • Focus • General purpose of the research • Outline of key issue • Writing arrangement
Summarize the important information of the above elements	Introduce the subject and foreground issues for discussion

（摘自　https://wenku.baidu.com/view/2dbcb868866fb84ae55c8d4d.html.）

第9章 文献检索

美国文献家赫伯特说:"知识的一半,是知道到哪里去寻找它。明日的文盲,不是不能阅读的人,而是缺乏检索能力的人。""工欲善其事,必先利其器",文献检索就是开展科学研究工作的有力武器。事实证明,任何一项知识创新、科学发明或学术成果的诞生,都需要查阅大量文献信息,借鉴和继承前人经验。因此,学会查找并阅读文献,对大学生来说特别重要,尤其是在当今互联网高速发展的时代。

文献检索(information retrieval)是指将信息按一定的方式组织和存储起来,并根据信息用户的需要找出有关的信息过程,所以它的全称又叫"信息的存储与检索"(information storage and retrieval),这是广义的信息检索。狭义的信息检索则仅指该过程的后半部分,即从信息集合中找出所需要的信息的过程,相当于人们通常所说的信息查询。

大学生能否熟练地检索和利用文献信息,能否借鉴和继承他人的经验和成果,是衡量其自学与知识更新能力、研究和开发能力、创新与突破能力的重要标志。面对浩如烟海的知识,最需要的是有用的专业知识,能够解决科学和教学实践中的知识。这样,就得学习和掌握如何在信息海洋中获取最需要的信息的本领——"沙海淘金"。利用完善的检索工具和检索系统,尤其是计算机的使用,将大大节省从业人员查阅文献的时间和精力,取得事半功倍的效果。

9.1 文献的类型

按照出版形式分,文献的类型主要有以下几种:

1. 图书

图书(book)是指对某一领域的知识进行系统阐述或对已有研究成果、技术、经验等进行归纳、概括的出版物。图书的内容比较系统、全面、成熟、可靠,但传统印刷业图书的出版周期较长,传递信息速度慢,电子图书的出版发行可弥补这一缺陷。

识别图书的主要依据有书名、著者、出版地、出版社、出版时间、总页数、国际标准书号(ISBN)等。ISBN由10位数字分成4段组成,各段依次是:地区或语种号-出版商代号-书名号-校验号。如7-302-02372-7,表示中国大陆代号为302的出版社(清华大学出版社),出版的一种图书,其书号为02372,该书的校验码为7。

图书在各种论文的参考文献或题录性检索工具中通常著录成如下例所示的格式:

Etten W V, 1991. Foundmentals of optical fiber communication[M]. London:Prentice-Hall.

2. 期刊

期刊(periodical/journal/serial)俗称杂志(magazine),是指有固定名称、版式和连续的编号,定期或不定期长期出版的连续性出版物。期刊是科技人员进行信息交流的正式、公开且有秩序的工具,被称为“整个科学史上最成功的无处不在的科学信息载体”。虽然全世界每年出版的期刊数量庞大,但是核心期刊数量有限,每个学科都有自己的核心期刊。专业核心期刊指刊载该专业论文数量较大(信息量较大)、学术水平较高的、能反映本学科最新研究成果及本学科前沿研究状况与发展趋势的、倍受该学科专业读者重视的期刊。核心期刊与非核心期刊是相对的,核心期刊是动态变化的。

识别期刊的主要依据有:期刊名称,期刊出版的年、卷、期,国际标准刊号(ISSN)等。ISSN由8位数字分两段组成,如1000-0135,前7位是期刊代号,末位是校验号。我国正式出版的期刊都有国内统一刊号(CN),它由地区号、报刊登记号和《中国图书馆图书分类法》分类号组成,如 CN 11-2257/G3。地区号依《中华人民共和国行政区划代码》(GB/2260-82)取前两位,例如,北京为11、天津为12、上海为31、辽宁为21、吉林为22等。

期刊论文的著录格式如下例所示:

[1] Tohyama H, 1991. A plasma Image bar for an electro-photographic printer[J]. Journal of the Imaging Science, 35:330-335.

3. 专利文献

专利文献(patent literature)是实行专利制度的国家,在接受申请和审批发明过程中形成的有关出版物的总称。包括专利说明书、专利公报、专利分类表、专利检索工具以及与此相关的法律性文件;狭义上专利文献仅指各国(地区)专利局出版的专利说明书或发明说明书。

专利公开号由国别代码(2位字母)+顺序号(7位数字)+法律状态码(1位字母)组成。如 US5489846A、CN1084635A、CN2302476Y。中国专利公开号、公告号的第一位数字表示专利类型(1为发明、2为实用新型、3为外观设计),后6位是流水号;法律状态码:A、C分别表示发明专利公开号、发明专利授权公告号,Y表示实用新型专利授权公告号,D表示外观设计专利授权公告号。中国专利申请号由9位数字组成:前两位表示申请年、第三位为专利类型、随后 5位是流水号、末位是校验码,如:013205838, 972087583。专利说明书常见的著录形式如下例所示:

Dayton B D. Differential amplifier apparatus: US,5095282,1992.

4. 标准文献

标准文献(standard literature)是经过公认的权威机构批准的以特定的文件形式出现的标准化工作成果。技术标准是对产品和工程建设质量、规格、技术要求、生产过程、工艺规范、检验方法和计量方法等所做的技术规定,是组织现代化生产、进行科学管理的具有法律约束力的重要文献。标准文献的特点是对标准化对象描述详细、完整、内容可靠、实用,有法律约束力,其时效性强,适用范围明确,是从事生产、设计、管理、产品检验、商品流通、科学研究的共同依据,也是执行技术政策所必需的工具。利用标准文献可了解有关方面的技术政策、生产水平和标准化水平，对引进、研制产品及设备，提高产品质量和生产水平，进行科学管理等有重要的参考价值。

标准文献都有标准号，它通常由国别（组织）代码+顺序号+年代组成，如ISO 3297-1986。我国的国家标准分为强制性的国标（GB）和推荐性的国标（GB/T），如 GB 18187-2000，GB/T2662-1999；行业标准代码以主管部门名称的汉语拼音声母表示，如 JT 表示交通行业标准；企业标准编号：Q/省、市简称+企业名代码+年份。识别标准文献的主要依据有标准级别、标准名称、标准号、审批机构、颁布时间、实施时间等。标准文献的常见著录形式如下：

BSI. Specification for communication and interference limits and measurements，BS，6839-1987.

5. 产品资料

产品资料（product literature）指产品目录、产品样本和产品说明书一类的厂商产品宣传和使用资料。产品样本通常对定型产品的性能、构造、用途、用法和操作规范等作具体说明，内容成熟，数据可靠，有的有外观照片和结构图，可直接用于在产品的设计制造中参考。产品技术资料著录的特点是：通常有表示产品样本一类资料的词，如 catalog、guide book、master of、data book of 等。产品技术资料的常见著录形式如下：

Integrated circuits Book IC11-Linear Products，Philips Data handbook of Philips Electronic Components and Materials Division，1988，p.3-131.

9.2 检索途径

检索途径也称检索入口，文献的特征是存储文献的依据，也是检索文献的依据，因此文献便构成了检索途径。

9.2.1 检索途径的类型

1. 外表途径

外表途径是指以文献外表特征标识作为检索入口的检索途径，主要包括以下几种类型：

（1）题名途径

题名途径（包括书名、篇名、刊名）指根据文献的名称来查找文献的途径，文献名主要指书名、刊名、会议名等，文献名索引按其名称顺序排列。这类检索工具有图书书名目录、期刊刊名目录等。

（2）著者途径

著者途径指根据著者的姓名来查找文献的途径。在检索工具中有著者索引和著者目录。著者索引按著者的姓或名字顺序排列后给出文摘号，著者索引又分为个人著者索引和团体著者索引。由于同一著者的文章往往具有一定的逻辑联系，以著者为线索可以系统、连续地掌握他们的研究水平和研究方向,因此具有族性检索的功能。

(3) 号码途径

号码途径指根据文献代码来查找文献的途径。号码有多种类型,如科技报告有报告号、标准文献有标准号、专利文献有专利号,使用这种序号索引查询很简单,但需要了解文献的号码,在事实检索中用得较多,一般情况下使用不多。

(4) 其他途径

其他途径有依据出版类型来检索的途径,有依据文献出版日期来检索的途径,有依据国别、文种来检索的途径。

2. 文献内容特征的检索途径

在不了解文献外表特征及并非确切查找某一指定文献而只是查找相关文献时,可使用文献内容特征的检索途径。

(1) 主题途径

主题途径是一种以代表文献主题内容的主题词为标引检索文献,通过主题目录和主题索引进行检索的方法。主题词按字母顺序排列,建立主题词索引(关键词索引、叙词索引),后面列出文摘号。

(2) 分类途径

分类途径以分类号作为检索标识,有分类目录,也有分类索引等。大多数检索工具的正文是按分类编排的,它们的目录就是分类索引。如大型工具书,书前有分类目录。分类目录或分类索引是以类名分类号作为检索标识。

(3) 其他途径

其他检索途径包括引文途径、分子式途径、化学物质途径,如分子式索引、环系索引、化学物质名称索引。

9.3　检索方法

1. 追溯法

追溯法又称回溯法,它是从已有的文献后面所附的参考文献入手,逐一追查原文,再从这些原文后面的参考文献逐一追查,不断丰富检索的线索,从而获得一批相关文献的查找方法。

追溯法的优点是在没有检索工具或检索工具不齐全的情况下,可以查到一批相关文献。而缺点是作者列的参考文献一般是有限的,有的与原文的关系不是很大,因此容易造成漏检和误检。

2. 常用法

常用法是指利用文摘、题录或索引等检索工具来查找文献的方法,也称为工具法。由于这种方法科学、省时、省力、效率高,是现在检索最常用的一种方法,所以称为常用法。按照所查文献的时间顺序，常用法分为以下3种：

(1) 顺查法

顺查法是以检索课题起始年代为起点，按时间顺序由远而近地查找文献的方法。查找前需了解课题的背景，通过有关的参考工具深入了解课题的实质性内容，再选择检索工具，从问题的发生年着手查起，查全率高，但工作量大，效率不高。一般在课题立项查新或成果鉴定查新时使用。

(2) 倒查法

倒查法是按逆时间顺序，由近而远地查找文献的方法。此法多用于新课题或老技术新发展的课题。课题对近期的状况比较重视，从新情况开始查找。

(3) 抽查法

抽查法是一种根据课题的特点和需要，选择该课题研究发展较快、出版文献较多的年代，抽取其中若干年，再进行查找的方法。该方法检索时间短，有漏检的可能，只有在对课题的学科发展前景有较多了解时才能使用。

3. 综合法

综合法以使用追溯法和常用法而得名，又称循环法或交替法。先利用检索工具查出一定时期内的一批有用文献，然后利用这些文献后的参考文献，再以追溯法查出前一时期内的文献，如此循环交替地使用上述两种方法，直到满足要求。

综合法兼有前两种方法的优点，全面而准确，适用于查阅那些过去年代内文献量较少的专业文献，并可弥补因检索工具不全而造成的漏检。

总之，以上各种方法各有长处和短处，实际检索文献时选用哪一种，要根据课题研究的需要以及所能利用的检索工具和检索手段。在检索工具较多的情况下，可以使用常用法；在已获得文献针对性很强的条件下，可利用追溯法获得相关性较强的文献；在获悉研究课题出版文献较多的年代即可利用抽查法。

9.4 检索步骤

1. 分析研究课题

分析检索课题所属学科范围明确检索的目的和意义，弄清课题的核心含义，确定所属学科、专业范围；检索时间范围根据课题的具体情况而定，课题查新就要检索近10年的文献，一般情况可长可短；分析文献的类型中现代科技文献种类很多，而不同类型的文献往往为不同的科技工作所需要。例如，从事一般性的科研、生产技术工作，需要掌握学科动态，主要利用期刊论文、会议文献、科技报告、专利文献；从事产品的设计和检验，则需要利用标准文献；从事发明创造，开展技术革新，需要检索专利文献；从事产品的外观设计，需要利用产品资料。

2. 选择检索工具

检索工具直接关系到检索效率和质量的高低。现在检索工具种类也很多,选择时可考虑以下几个方面的因素:① 检索工具报道文献的学科专业范围;② 检索工具所报道的文献类型;③ 检索工具所收录文献的语种;④ 检索工具提供的检索途径,同时要从现有的检索工具情况出发。如果对本专业有哪些检索工具或检索工具报道的专业范围不太清楚,可利用介绍检索工具的工具书,如中国科学技术情报研究所编著的《国外科技文献检索工具书简介》及其续编,辽宁人民出版社出版的《科技检索工具书综录》(上、下册)等。

3. 确定检索途径和检索策略

在利用检索工具查找文献时,主要是利用检索工具的各种索引,即通过各种检索途径来查找文献线索。一般来说每种检索工具都提供了多条检索途径。如果检索课题要求的是泛指性较强的文献信息,则最好选择分类途径,如果要求专指性较强的文献信息,最好选择主题途径,如果事先已知文献题名、著者、号码(专利号、标准号、报告号)等条件,可利用相应的各种途径。但也要依据选择的具体检索工具而定,有什么途径用什么途径。检索策略就是提取检索词并确定组配关系。在检索过程中,检索策略可能要经过多次改进,如检索文献太多,就要缩小检索范围。

4. 选择检索方法

依据所要检索的内容具体选择何种检索方法可参见9.2节,此处不再赘述。

5. 查找和获取原文

根据所选定的检索途径和方法,通过检索工具的各种索引查得文献顺序后,进而查文献或题录,获得文献出处,最后获取原文。

首先,判断文献的出版类型。其次,整理文献出处。检索工具中文献出处通常采用缩写,有音译刊名时,需要通过“来源索引”“收录出版物一览表”等还原成全称或原名。最后,根据出版类型在图书馆或情报所查找馆藏目录或联合目录确定是否收藏。

在检索工具中,不管是题录还是文摘,对摘录的文献的类型不进行明显区分,需要我们自己辨识。

9.5 《工程索引》对英文摘要的基本要求

《工程索引》要求信息性文摘(information abstract)应该用简洁、明确的语言(一般不超过 150 个单词)将论文的目的(purposes)、主要的研究过程(procedure)及所采用的方法(methods),由此得到的主要结果(results)和得出的重要结论(conclusions)表达清楚。如有可能,还应尽量以一句话概括论文结果和结论的应用范围和应用情况。也就是说,要写好英文摘要,作者必须回答好以下几个问题。

1. 本文的目的或要解决的问题

主要说明作者写作此文的目的,或本文主要解决的问题。一般来说,一篇好的英文摘要,开头就应该把作者写本文的目的或要解决的主要问题非常明确地交代清楚。必要时,可利用论文中所列的最新文献,简要介绍前人的工作,但这种介绍一定要极其简练。不谈或尽量少谈背景信息,避免在摘要的第一句话重复使用题目或题目的一部分。

2. 解决问题的方法及过程

主要说明作者主要工作过程及所用的方法,也应包括众多的边界条件,使用的主要设备和仪器。在英文摘要中,过程与方法的阐述起着承前启后的作用。开头交代了要解决的问题(What I want to do)之后,接着要回答的自然就是如何解决问题(How I did it),而且,最后的结果和结论也往往与研究过程及方法是密切相关的。大多数作者在阐述过程与方法时,最常见的问题是泛泛而谈、空洞无物,只有定性的描述,使读者很难清楚地了解论文中解决问题的过程和方法。因此,在说明过程与方法时,应结合(指向)论文中的公式、实验框图等进行阐述,这样既可以给读者一个清晰的思路,又可以给那些看不懂中文(但却可以看懂公式、图、表等)的英文读者以一种可信的感觉。

3. 结果和结论

结果和结论部分代表着文章的主要成就和贡献,论文有没有价值,值不值得读者阅读,主要取决于你所获得的结果和所得出的结论。因此,在写作结果和结论部分时,一般都要尽量结合实验结果或仿真结果的图、表、曲线等来加以说明,使结论部分言之有物,有根有据;同时,对那些看不懂中文的英文读者来说,通过这些图表,结合英文摘要的说明就可以比较清楚地了解论文的结果和结论。也只有这样,论文的结论才有说服力。特别地,在结尾部分还可以将论文的结果和他人最新的研究结果进行比较,以突出论文的主要贡献和创新、独到之处(回答“What is new and original in this paper?”)。

为了提高文字效能,应尽量删去多余的字句。摘要中只谈新的信息,尽量使摘要简洁。

9.6 常用网络数据库简介

1. 万方数据资源系统

万方数据资源系统是1997年8月由中国科技信息研究所、万方数据集团公司联合开发的网上数据库联机检索系统。万方数据知识服务平台整合数亿条全球优质知识资源,集成期刊、学位、会议、科技报告、专利、标准、科技成果、法规、地方杂志、视频等十余种知识资源类型,覆盖自然科学、工程技术、医药卫生、农业科学、哲学政法、社会科学、科教文艺等全学科领域,实现海量学术文献统一发现及分析,支持多维度组合检索,适合不同用户群使用。万方智搜致力于“感知用户学术背景,智慧你的搜索”,帮助用户精准发现、获取与沉淀知识精华。期刊资源包括国内期刊和国外期刊,其中国内期刊共8000余种,涵盖自然科学、工程技术、医药卫生、农业科学、哲学政法、社会科学、科教文艺等多个学科;国外期刊共包含40000余种世界各国出版的重要学术期刊,主要来源于国家科技图书文献中心(NSTL)外文文献数据库以及数十家著名学术出版机构。

2. 中国知网(CNKI)

CNKI是中国知识基础设施工程(China National Knowledge Infrastructure)的简称,是综合性的大型数据库,包含中国期刊全文数据库,中国优秀硕、博士学位论文全文数据库,中国重要会议论文全文数据库,中国重要报纸全文数据库,中国统计年鉴数据库,中国精品文艺作品期刊文献库,中国法律知识资源总库等多个数据库,覆盖的学科范围包括数理科学、化学化工和能源与材料、工业技术、农业、医药卫生、文史哲、经济政治与法律、教育与社会科学、电子技术与信息科学等 。

3. Springer Link电子期刊网络数据库

Springer Link是全球最大的在线科学、技术和医学(STM)领域学术资源平台。凭借弹性的订阅模式、可靠的网络基础以及便捷的管理系统,Springer Link已成为各家图书馆最受欢迎的网站。通过Springer Link的IP网址,读者可以快速地获取重要的在线研究资料。Springer Link提供多种远端存取方式,包括通过IP认证、Athens或Shibboleth等认证方式。Springer已经出版超过150位诺贝尔奖得主的著作。目前,Springer Link正为全世界600家企业客户、超过35000个机构提供服务。Springer Link的服务范围涵盖各个研究领域,提供超过1900种同行评议的学术期刊,以及不断扩展的电子参考工具书、电子图书、实验室指南、在线回溯数据库以及更多内容。

【练习9.1】 利用Springer Link数据库,查出相关课题文献,将结果填写在表9.1中。

检索课题:化工原理。

检索主题(英文):Chemical　principle。

检索要求:检索式中至少包含两个以上检索词,检索结果命中文献在100篇以内。

Table 9.1　Search content

表9.1　检索内容

所选字段		检索词
1	TI　Title	chemical
2	TI　Title	principle
3		
检索式:TI chemical principle and TI principle		
限定条件	出版日期	2018~2020
	其他条件	
检索结果(命中文献篇数)________,任选其中一篇,填写相关内容如下:		
Title		
Author		
Source		

【练习9.2】 利用"Advanced Search"查出相关课题文献,将结果汇总在表9.2中。

检索结果:最终检出文献________篇。选择其中一篇期刊论文,摘录如下:

Table 9.2 Search content

表9.2 检索内容

Title	
Author	
Source(Journal, Issue)	

【练习9.3】 选择检索结果中的任意一篇文献,将相关内容填入表9.3中。

Table 9.3 Search content

表9.3 检索内容

Title	
Author	
Source	
Abstract	

材料阅读

How to Write a Research Paper?

Step 1. Choosing a Topic

(1) Selecting a general topic.

(2) Reading and thinking.

(3) Narrowing down the scope of your topic to a facet or facets which can be developed into a research paper.

(4) Formulating the final topic.

- Serious, meaningful.
- Within your ability.
- Sufficient materials are available.
- Objective, not subjective, "which was the greater poet, Libai or Dufu?", "A comparative study of the themes in Libai's and Dufu'spoem".
- Important guideline in choosing your topic: English language(including translation), literature in English, society and culture of countries where English is spoken.

Step 2. Searching for Sources

(1) When you get a book, you should skim for major ideas first.

- Glance at the preface of a book.
- Look up the subject in the index of the book.
- Read the chapter headings.
- Read the first and last two sentences in paragraph to find out what information it contains.
- Glance at the opening paragraph of an article, essay, or book chapter.
- Glance at concluding paragraphs in an article, essay, or book chapter.
- Run your eye down the page, reading randomly every fourth or fifth sentence.

(2) Evaluating sources.

(3) Compiling the working bibliography.

You should record the following information:

- Name of the author, last name first.
- Title of the book, underlined.
- Place of publication.
- Publisher's name.
- Date of publication.

Step 3. Reading the Sources and Taking Notes

(1) To record the general ideas that will form the backbone of our research paper.

(2) To record specific pieces of information that support the general ideas.

(3) To preserve the exact wording of some statements in your sources that you may want to quote directly in the paper.

Step 4. Thesis Statement and Outline

The thesis is a statement that summarizes the central idea of the paper. It is usually the final sentence of the opening paragraph.

Rules for wording the thesis:

(1) The thesis should commit the writer to a single line of argument.

(2) The thesis should not be worded in figurative language.

(3) The thesis should not be vaguely worded.

(4) The thesis should not be worded as a question.

(5) The thesis should be as concise as possible.

Step 5. Quoting and Citing Skills

There are two kinds of quotation:

(1) A direct quotation, where you use the writer's actual words.

(2) An indirect quotation, where you summarize the writer's ideas and put them into your own words.

Step 6. Revising the Rough Draft

Once you have brought your outline to life by writing a rough essay, the nature of your job changes significantly. You must now become your own toughest critic. Then you can become the author again, rewriting and, if necessary, reorganizing the weaker passages so that they become as strong as the best. Finally, you must examine the revised essay very closely in order to correct the spelling, punctuation, and other mechanical details.

(摘自 http://www.wendangku.net/doc/860f9f6c0066f5335b81211f.htmlpiling the working bibliographg.)

第10章　文献综述的写作

文献综述(literautre review)是科研论文中的重要组成文体之一。它以作者对各种文献资料的整理、归纳、分析和比较为基础,就某个专题的历史背景、前人的工作、研究现状、争论的焦点及发展前景等方面进行综合、总结和评论,反映了当前某一领域中某分支学科或重要专题的最新进展、学术见解和建议。因此,它往往能反映出有关问题的新动态、新趋势、新水平、新原理和新技术等。文献综述往往被收集在专业期刊的Review栏目中,常见的有Survey、Advances、Progress、Recent Advances、Update 和Annual Review等。

根据写作目的和内容的侧重点,文献综述大致可以分为动态性综述(developmental review)、成就性综述(result review)和争鸣性综述(contentious review)。按时间来划分,文献综述又可分为回顾性综述(retrospective review)和前瞻性综述(prospective review)。根据作者的参与情况,文献综述还可分为归纳性综述(inductive review)和评论性综述(critical review)。文献综述的分类并非绝对,在实际写作中,往往是各种类型综合在一起。同学们在写作综述的时候,至少有以下目的:

(1) 通过搜集文献资料过程,可进一步熟悉科学文献的查找方法和资料的积累方法;在查找的过程中同时也扩大了知识面。

(2) 查找文献资料、写文献综述是科研选题及进行科研的第一步,因此学习文献综述的撰写也是为今后科研活动打基础的过程。

(3) 通过综述的写作过程,能提高归纳、分析、综合能力,有利于独立工作能力和科研能力的提高。

(4) 文献综述选题范围广,题目可大可小,可难可易。对于毕业设计的课题综述,则要结合课题的性质进行书写。

一篇结构完整的文献综述应由六个部分组成:标题与作者(title and author)、摘要与关键词(abstract and key words)、引言(introduction)、述评(review)、结论(conclusion)和参考文献(references)。与研究性论文相比,文献综述的篇章结构比较自由,但其中最为核心的部分是引言、述评和结论。

10.1　引　　言

引言是文献综述正文的开始部分,主要包括两个方面:一是提出问题;二是介绍综述的范围和内容。提出问题时,作者要给出定义性解释、交代研究背景、简单介绍不同文献的看法和分歧所在,并介绍该文献的写作目的;介绍该综述的范围和主要内容时,作者应使用简明扼要的语句加以概括。引言部分通常为200~300词。

引言的内容和结构具有以下特点:

(1) 综述的引言通常包括定义性解释,研究背景,现存问题或分歧,综述的目的、内容和范围。

(2) 使用一般现在时介绍背景知识,使用现在完成时叙述他人成果,使用一般将来时或一般现在时简介本文内容。

(3) 句子结构力求简洁明了,多用简单句,并列成分较为常见。

(4) 以第三人称主语为主,间或使用第一人称复数充当主语。

要将引言的内容清楚地用英文表述出来,常常需要借助以下的语言形式。

(1) 表定义或解释:

① ……被定义为…… ...is defined as...; ...has been defined as...

② 所谓……是指…… By...is meant...; By...we mean...

③ ……指的是…… ...refers to...

④ ……包括…… ...includes...

⑤ ……有……种类型 ...is / can be classified into...; There are;...kinds of...

⑥ 我们知道,……是一种……的常见病 ...is a common disease that...; ...is known to be a common disease that...

(2) 表现状和分歧:

① 据发现/ 报道/ 证实…… It has been found/reported/proved that...

② 普遍认为…… It is generally recognized / agreed/ accepted that...

③ 一般认为…… It is thought / regarded/ considered that...

④ ……依然是一个尚待解决的问题 ...remains an unsolved problem.

⑤ 关于……目前有两种解释 Two theories have been postulated to explain...

⑥ 第一种理论认为……,而第二种理论则认为…… The first theory proposes that...; whereas the second theory proposes that...

⑦ 一些文献报道……而另一些人持不同观点 Some papers have reported that... however, other groups have disputed these findings.

⑧ 最初的一些研究支持这种看法 Several initial studies seemed to support this concept.

(3) 表内容和目的:

① 本综述的目的是…… The purpose / aim / object of this reviewis to...

② 本文综述了有关文献 The pertinent literature is reviewed.

③ 本文综述了…… This article reviews.../We review...

④ 本文将重点讨论…… This review will concentrate on...

⑤ 下面本文就……作一简单综述 In the following, a brief review is given of / about...

⑥ 本篇综述的目的是着重阐述……　In this review, we aim to highlight...

⑦ 我们将回顾有关……的研究　We will review published studies on...

⑧ 我们将重点回顾……　We will focus on...

⑨ 本文主要阐述……　This review focuses on...

⑩ 本文就……作一综述　The following paper reviews...

⑪ 本文并非旨在说服读者……　No attempt will be made to convince the reader that...

10.2　述　　评

述评是文献综述的核心所在，是对引言的展开和深入。根据引言所提出的问题和限定的范围，作者要对大量有关文献进行系统的整理、归纳、对比和分析，在此基础上列出与主题有关的所有重要学术观点，然后分别加以论述，以便读者获得全面的了解。动态性综述在论述观点时，通常遵循由一般到具体、由过去到最近、由他人观点到自己看法的顺序。引述文献时，只介绍主要研究成果和结论性意见，对于研究的材料、方法和过程则不必详述，但成就性综述在介绍创新点时则应多加论述。对于争鸣性综述中尚无定论和存在分歧的观点，只需归纳提及，由读者自己进行思考和判断。述评部分通常较长，为了条理清晰，作者一般将其分成几个部分并给出每个部分的小标题。述评的内容和结构常具有以下特点：述评部分常由几个部分组成，每个部分又有各自的标题及下级标题。回顾前人研究，以时间为序，由远及近。以有叙有议的方式体现述评的功能，叙前人研究，议其结果、探其原因、究其不足。

叙述时使用一般过去时，评论时使用一般现在时或现在完成时。分析评论，特别是表达作者自己的观点时，要客观、谨慎，因此多使用模糊性语言和表推测的语言形式。述评是文献综述的主体部分，借助符合科技文体写作规范的语言形式就显得颇为重要。善于使用下列表达不仅能使文章流畅清晰，还能使文章彰显学术性。

（1）表述观点：

① 研究表明……　Studies show that...

② 据(已经)发现……　It is (has been) found that...; Found that...

③ 据报道……　It is (has been) reported that...

④ 有人指出……　It has been pointed out that...

⑤ 业已证明……　It has been proved / showed that...

⑥ 一般认为……　It is generally recognized / agreed / accepted that...

⑦ 有人认为……　It is thought / regarded / considered that...

⑧ 已观察到……　It has been observed that...

⑨ 必须指出……　It must be pointed out that...

⑩ 还得指出……　It should be added that...

⑪ 必须承认……　It must be admitted that...

⑫ 不用说…… It need not be said that...

⑬ 必须强调…… It must be emphasized / stressed that...

⑭ 应当讲明…… It should be made clear that...

⑮ 一项有趣的发现是…… An interesting finding is that...

⑯ 最重要的事实是…… Nothing is more important than the fact that...

⑰ 更重要的事实是…… A more important fact is that...

⑱ 我认为…… I am of the opinion that...

⑲ 有人声称…… It is asserted that...

⑳ 多数人一致认为…… Most researchers agree that...

(2) 探讨可能性:

① 可以有把握地说…… It may be safely said that...

② 由此可见…… It can be seen from this that...

③ ……有(不)可能 It is (not) possible / probable / likely that...

④ ……是合乎情理的 It stands to reason that...

⑤ 毫无疑问 There is no doubt that...

⑥ 显然…… It is obvious / clear / apparent / evident that...

⑦ 目前尚无临床证据说明…… There is no clinical proof of ...

(3) 表比较和对照:

① 使用句型。

(a) A 与 B 之间存在差异(相似点):There are some / a few / many differences (similarities) between A and B.

(b) A 与 B 在三个方面有不同点:A differs from B / is different from B in three aspects.

(c) 一个不同(相同)之处在于:One difference (similarity) seems to be / lies in / is that...

② 词及词组。常用的词及词组有 on the contrary、in contrast、in comparison、on the other hand、be like (unlike)、just as、be the same as、similarly、likewise、while、whereas、yet、but、however、differently。

10.3 结 论

结论不仅是作者对全文的总结,也是作者发表个人意见的部分,一般有标题 conclusion 或 summary,较短的综述如果没有小标题,则往往有 as mentioned above、to sum up、to conclude、inshort、in all 等短语引出结论。结论的内容包括:对述评的归纳、对各种问题的评论性意见、对未来研究的建议或展望。对述评的归纳还可以逐条加序号撰写。有的文献综述不在最后作总体的结论,而是放在述评各部分之后分别总结,内容分别与述评的几大部分相对应,但笔墨的重点放在结果的陈述与评价上。每个部分又由事实陈述、结果评价和研究预测组成。结论是述评的浓缩,语言方面亦如此,除了使用一般现在时态外,大量使用现在完成时,以强调该研究到目前所取得的成果。另外,比较和评价的语言形式也是该部分的语言特点之一。要写好英语文献综述并非一朝一夕的易事,在对其内容、结构和语言的特征有所了解后,持之以恒地写和练一定会提高写作水平。

首先,需要将"文献综述"(literature review)与"背景描述"(backupground description)区分开来。"文献综述"是对学术观点和理论方法的整理。

其次,文献综述是评论性的(Review 就是"评论"的意思),因此要带着作者本人批判的眼光(critical thinking)来归纳和评论文献,而不仅仅是相关领域学术研究的"堆砌"。评论的主线,要按照问题展开,也就是说,别的学者是如何看待和解决你提出的问题的,他们的方法和理论是否有什么缺陷?要是别的学者已经很完美地解决了你提出的问题,那就没有重复研究的必要了。

文献综述是对选题阅读工作的进一步整理,其写作技巧主要包括以下几个方面:

(1) 瞄准主流。主流文献,如该领域的核心期刊、经典著作、专职部门的研究报告、重要的观点和论述等,是写文献综述的"必修课"。而多数大众媒体上的相关报道或言论,虽然有点价值,但时间精力有限,可以从简。

(2) 随时整理,如对文献进行分类,记录文献信息和藏书地点。写论文的时间很长,有的文献看过了当时不一定有用,事后想起来却找不着了,所以有时记录是很有必要的。对于特别重要的文献,不妨做一个读书笔记,摘录其中的重要观点和论述。这样一步一个脚印,到真正开始写论文时就积累了大量"干货",可以随时使用。

(3) 要按照问题来组织文献综述。有学者说:"文献综述就像是在文献的丛林中开辟道路,这条道路本来就是要指向我们所要解决的问题,当然是直线距离最短、最省事,但是一路上风景颇多,迷恋风景的人便往往绕行于迤逦的丛林中,反而'乱花渐欲迷人眼',找不到问题主线了。"因此,在写文献综述时,头脑要时刻清醒:我要解决什么问题,人家是怎么解决问题的,说得有没有道理。

10.4　文献综述范文格式

一篇完整的应用/研究综述要包括以下几个方面的内容:

(1) 作者姓名、作者单位、通讯地址,摘要(abstract: 200~500 词),关键词(key words: 5个词以内)。

(2) 引言(introduction)。进行该项研究的意义(或者为什么要开展该项研究 / 为什么要写这篇文章)。

(3) 国内外研究进展与现状(literature review)。在国内外,都有哪些人或者哪些机构在该领域做了哪些事,开展了哪些方面的研究,他们采用的是什么方法,建立了什么模型,使用了什么软件,解决了什么问题,优势是什么,不足又是什么,对于他们存在的不足,你有什么可能的解决方法。

(4) 未来展望(future work)。对于目前国内外的研究现状,你认为未来可从哪些方面加以深入研究。

(5) 参考文献(references)。

(6) 英文标题(包括作者姓名、单位和通讯地址)、英文摘要、英文关键词。

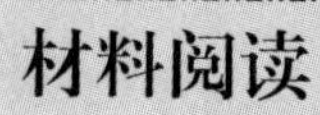

How to Write a Literature Review ?

1. Definition

A literature review is both a summary and explanation of the complete and current state of knowledge on a limited topic as found in academic books and journal articles. There are two kinds of literature reviews you might write at university: one that students are asked to write as a stand-alone assignment in a course, often as part of their training in the research processes in their field, and the other that is written as part of an introduction to, or preparation for, a longer work, usually a thesis or research report. The focus and perspective of your review and the kind of hypothesis or thesis argument you make will be determined by what kind of review you are writing. One way to understand the differences between these two types is to read published literature reviews or the first chapters of theses and dissertations in your own subject area. Analyze the structure of their arguments and note the way they address the issues.

2. Purpose of the Literature Review

It gives readers easy access to research on a particular topic by selecting high quality articles or studies that are relevant, meaningful, important and valid and summarizing them into one complete report. It provides an excellent starting point for researchers beginning to do research in a new area by forcing them to summarize, evaluate, and compare original research in that specific area.

It ensures that researchers do not duplicate work that has already been done.what's more, it can provide clues as to where future research is heading or recommend areas on which to focus, and it highlights key findings. Moreover, it identifies inconsistencies, gaps and contradictions in the literature.

3. Content of the Review

(1) Introduction

The introduction explains the focus and establishes the importance of the subject. It discusses what kind of work has been done on the topic and identifies any controversies within the field or any recent research which has raised questions about earlier assumptions. It may provide background or history. It concludes with a purpose or thesis statement. In a stand-alone literature review, this statement will sum up and evaluate the state of the art in this field of research; in a review that is an introduction or preparatory to a thesis or research report, it will suggest how the review findings will lead to the research the writer proposes to undertake. It provides a constructive analysis of the methodologies and approaches of other researchers.

(2) Body

Often divided by headings/subheadings, the body summarizes and evaluates the current state of knowledge in the field. It notes major themes or topics, the most important trends, and any findings about which researchers agree or disagree. If the review is preliminary to your own thesis or research project, its purpose is to make an argument that will justify your proposed research. Therefore, it will discuss only that research which leads directly to your own project.

(3) Conclusion

The conclusion summarizes all the evidence presented and shows its significance. If the review is an introduction to your own research, it highlights gaps and indicates how previous research leads to your own research project and chosen methodology. If the review is a stand-alone assignment for a course, it should suggest any practical applications of the research as well as the implications and possibilities for future research.

4. Nine Steps to Writing a Literature Review

(1) Find a working topic.

Look at your specific area of study. Think about what interests you, and what fertile ground for study is. Talk to your professor, brainstorm, and read lecture notes and recent issues of periodicals in the field.

(2) Review the literature.

• Using keywords search a computer database. It is best to use at least two databases relevant to your discipline.

• Remember that the reference lists of recent articles and reviews can lead to valuable papers.

• Make certain that you also include any studies contrary to your point of view.

(3) Focus your topic narrowly and select papers accordingly.

Consider the following:

• What interests you?

• What interests others?

• What time span of research will you consider?

Choose an area of research that is due for a review.

(4) Read the selected articles thoroughly and evaluate them.

• What assumptions do most/some researchers seem to be making?

• What testing procedures, subjects, material tested?

• Evaluate and synthesize the research findings and conclusions drawn.

• Note experts in the field: names/labs that are frequently referenced.

• Note conflicting theories, results, and methodologies.

• Watch for popularity of theories and how this has/has not changed over time.

(5) Organize the selected papers by looking for patterns and by developing sub-topics.

Note things such as:

• Findings that are common/contested...

• Two or three important trends in the research...

• The most influential theories...

(6) Develop a working thesis.

Writting a one or two sentence statement summarizing the conclusion you have reached about the major trends and developments you see in the research that has been done on your subject.

(7) Organize your own paper based on the findings from steps 4 and 5.

• Develop headings/subheadings.

If your literature review is extensive, find a large table surface, and on it place post-it notes or filing cards to organize all your findings into categories. Move them around if you decide that (a) they fit better under different headings, or (b) you need to establish new topic headings.

(摘自 https://www.learning commons.uoguelph.cn.)

附　　录

附录1　希腊字母英文名称

大写	小写	英文名
A	a	alpha
B	b	beta
G	g	gamma
D	d	delta
E	e	epsilon
Z	z	zeta
H	h	eta
Q	q	theta
I	i	iota
K	k	kappa
L	l	lambda
M	m	mu
N	n	nu
X	x	xi
O	o	omicron
P	p	pi
R	r	rho
S	s	sigma
T	t	tau
U	u	upsilon
F	f	phi
C	c	chi
Y	y	psi
W	w	omega

附录2　部分常用数学符号及数学式的读法

1. 小数的读法

小数部分的表达需分别读出每个数，小数点读作point。例如：

小数	英文
3.576	three point five seven six
0.45	zero point four five; point four five
13.91	thirteen decimal (point) nine one
0.23	zero point two three
$5.50\dot{1}\dot{4}$	five point five zero one four, one four recurring

2. 算术式的读法

符号或算术式	英文
+	plus; positive
−	minus; negative
±	plus ; minus
×	multiplied by; times
÷	divided by
=	is equal to; equals
≡	is identically equal to
≈	is approximately equal to
2+3=5	Two plus three is(equals,is equal to)five
5−3=2	Five minus three is equal to two
3×2=6	Three times two is six; Three by two are six
9÷3=3	Nine divided by three makes three

3. 温度的读法

例如：

100 °C：one hundred degrees Centigrade[Celsius].

32 °F：thirty-two degrees Fahrenheit.

4. 百分数和千分数(percent and permille)的读法

例如：

6%：6 percent; six hundredths.

3‰：3 permille; three thousandths.

5. 整数的读法

整数	英文读法
1	one
10	ten
100	one hundred
1000	one thousand
10000	ten thousand
100000	one hundred thousand
1000000	one million
3004012	three million, four hundred (and) twelve
12020003	twelve million, twenty thousand (and) three
9192631770	nine billion, one hundred and ninety two million, six hundred and thirty one thousand, seven hundred and seventy
1000000000	one billion
1000000000000	one trillion

6. 其他常见数学符号的读法

数学符号	英文读法
∞	infinite
b'	b prime
b''	b double prime; b second prime
b'''	b triple prime
A_1	A sub one
A_n	A sub n
b_m^n	b double prime sub m
$x!$	factorial x
x^n	the nth power of x
$x^{1/n}$	the nth root of x; x to the power one over n
®	approach; tend to
$\Rightarrow$	implies
$\Leftrightarrow$	is equivalent to
$\neq$	is not; is not equal to
$<$	is less than
$>$	is greater than
$\leqslant$	is less than or equal to
$\geqslant$	is greater than or equal to
$\in$	is a member of set
$\frac{\mathrm{d}y}{\mathrm{d}x}$	the first derivative of y with respect to x

续表

数学符号	英文读法
$\frac{\partial y}{\partial u}$	the partial derivative of y with respect to u, where y is a function of u and another variable(or variables)
$\int_a^b$	integral between limits a and b

附录3　化学式的英文读法

1. 化学符号的读法

化学符号	英文读法
+	plus
-	minus
( )	round brackets; brackets; parentheses; in brackets
(	open bracket
)	close bracket
[ ]	square brackets
=	equals; is equal to
$\rightleftharpoons$	reacts reversibly
→	give(s); yield(s)
↑	evolved as a gas
↓	is precipitated; precipitates; gives an precipitate
·	dot
$\xrightarrow{Cu,\triangle}$	in the presence of copper as a catalyst on heating, give(s)
$Ca_2(PO_4)_3 \cdot 2H_2O$	calcium phosphate two hydrate; C-a-two-pause-P-O four-pause-three times-dot-two-H-two-O
$[Zn(NH_3)_4]^{2+}$	tetra-ammonium zinc complex cation
$CO_3^{2-} + Ca^{2+} = CaCO_3$	a carbonate anion with a valency of two plus a calcium cation with a valency of two produce a calcium carbonate precipitate

2. 其他

化学式中的英语字母，不论大小写，一律读其在字母表中的读音，如A读 [ei]，b读 [biː]。阿拉伯数字1, 2, 3, 4…直接读one, two, three, four……。圆括号右下标的数字则读成twice, three times, four times……。

参考文献

[1] 《化工英语》编写组. 化工英语[M]. 北京: 高等教育出版社,2017.

[2] 杨春华,陈刚. 化工专业英语[M]. 北京: 化学工业出版社,2011.

[3] 符德学. 化学化工专业英语[M]. 北京: 化学工业出版社,2011.

[4] 刘庆文. 化工专业英语[M]. 北京: 化学工业出版社,2010.

[5] 刘宇红. 化学化工专业英语[M]. 北京: 中国轻工业出版社,2001.

[6] 邵荣,许伟, 吕慧华. 新编化工专业英语[M]. 2版. 上海: 华东理工大学出版社,2017.

[7] 胡鸣,刘霞. 化学工程与工艺专业英语[M]. 北京:化学工业出版社,2007.